Matthias Templ

New Developments in Statistical Disclosure Control and Imputation

Matthias Templ

New Developments in Statistical Disclosure Control and Imputation

Robust Statistics Applied to Official Statistics

Südwestdeutscher Verlag für Hochschulschriften

Impressum/Imprint (nur für Deutschland/ only for Germany)
Bibliografische Information der Deutschen Nationalbibliothek: Die Deutsche Nationalbibliothek verzeichnet diese Publikation in der Deutschen Nationalbibliografie; detaillierte bibliografische Daten sind im Internet über http://dnb.d-nb.de abrufbar.
Alle in diesem Buch genannten Marken und Produktnamen unterliegen warenzeichen-, marken- oder patentrechtlichem Schutz bzw. sind Warenzeichen oder eingetragene Warenzeichen der jeweiligen Inhaber. Die Wiedergabe von Marken, Produktnamen, Gebrauchsnamen, Handelsnamen, Warenbezeichnungen u.s.w. in diesem Werk berechtigt auch ohne besondere Kennzeichnung nicht zu der Annahme, dass solche Namen im Sinne der Warenzeichen- und Markenschutzgesetzgebung als frei zu betrachten wären und daher von jedermann benutzt werden dürften.

Verlag: Südwestdeutscher Verlag für Hochschulschriften Aktiengesellschaft & Co. KG
Dudweiler Landstr. 99, 66123 Saarbrücken, Deutschland
Telefon +49 681 37 20 271-1, Telefax +49 681 37 20 271-0, Email: info@svh-verlag.de
Zugl.: Wien, TU, Diss., 2009

Herstellung in Deutschland:
Schaltungsdienst Lange o.H.G., Berlin
Books on Demand GmbH, Norderstedt
Reha GmbH, Saarbrücken
Amazon Distribution GmbH, Leipzig
ISBN: 978-3-8381-0828-5

Imprint (only for USA, GB)
Bibliographic information published by the Deutsche Nationalbibliothek: The Deutsche Nationalbibliothek lists this publication in the Deutsche Nationalbibliografie; detailed bibliographic data are available in the Internet at http://dnb.d-nb.de.
Any brand names and product names mentioned in this book are subject to trademark, brand or patent protection and are trademarks or registered trademarks of their respective holders. The use of brand names, product names, common names, trade names, product descriptions etc. even without a particular marking in this works is in no way to be construed to mean that such names may be regarded as unrestricted in respect of trademark and brand protection legislation and could thus be used by anyone.

Publisher:
Südwestdeutscher Verlag für Hochschulschriften Aktiengesellschaft & Co. KG
Dudweiler Landstr. 99, 66123 Saarbrücken, Germany
Phone +49 681 37 20 271-1, Fax +49 681 37 20 271-0, Email: info@svh-verlag.de

Copyright © 2009 by the author and Südwestdeutscher Verlag für Hochschulschriften Aktiengesellschaft & Co. KG and licensors
All rights reserved. Saarbrücken 2009

Printed in the U.S.A.
Printed in the U.K. by (see last page)
ISBN: 978-3-8381-0828-5

Summary

The terminology *Official Statistics* basically refers to those fields of statistics which are practised by official statistical institutions. Data handled by these institutions is special in a particular way. Samples are taken from finite populations about which (rare) information is available. This kind of information is continuously updated in registers.

While official statistical institutions hold a monopoly in providing official statistics, statistical methods for creating these statistics are not developed by official statistical institutions only. Numerous research institutes and universities work in the various fields of official statistics: detecting errors in the data (*Editing*), imputation of missing values, forecasting and seasonal adjustment in time series, linking data sets with non-explicit key values (*Statistical Matching*), graphical processing of data, estimation of indicators, estimating statistics for small areas (*Small Area Estimation*) and privacy of statistical data (*Statistical Disclosure Control*).

Especially the problem of privacy of statistical data has gained tremendous significance over the last two decades. Official laws, particularly the data protection laws and the Federal Statistics Law prohibit any re-identification of statistical units (e.g. persons or businesses) after the transfer of data to public or research institutions.

During the last few years there has been an increasing demand for data among researchers. Because of this rising demand for microdata (unaggregated single data sets) statistical disclosure control (SDC) has developed into a large field of research. Its aim is to keep up the required statistical privacy while making viable data available to the researchers. This can be achieved with the help of minimal modifications of the data without changing the multivariate data structure. Research in the field of statistical data disclosure has particularly increased in Statistik Austria in connection with this book [see Templ, 2008a].

Statistical institutions make their data available to the public at least in an aggregated form. These data is often processed in hierarchical tables (containing margins) in publications or by web applications. Here no statistical interference to single statistical units shall be possible, hence particular values are suppressed or modified in the tables. In order to perform as little suppression or modification as possible and in order to keep up the data protection at the same time, highly-complex mathematical methods have to be applied in order to solve large integer linear programs with thousands of constraints.

The right handling of missing values is essential in the field of data disclosure control, but in numerous other fields of official statistics as well. Missing values in official statistics are the result of unanswered questions in questionnaires. But also obviously incorrect entries in the data sets are changed to missing. These missing values must then be estimated (Data Imputation). The quality of the imputation defines the quality of the data to a great extent and eventually the quality of the indicators, too, which is important for politics and economy. On the other hand, the choice of the (right) imputation method depends on the data in question and on the distribution of the missing values.

In order to create some awareness for this problem within statistical institutions special training of the employees is essential (especially in connection with computer applications and theory) [Dinges and Templ, 2009].

Chapter 1 contains a small introduction into methods relevant for this book as far as they are not explained in the successive chapters. Almost all these chapters are composed of articles already published in established and re-viewed journals or which have been accepted for publication.

Chapter 2 introduces the well-developed R package *sdcMicro* [Templ, 2007c, 2008d]. This package is freely available and can be obtained through the *Comprehensive R Archive Network* (CRAN), (http://cran.r-project.org). With the help of this package it is possible to keep microdata confidential in a very effective way. The concept and the structure are thoroughly explained and its application is demonstrated using data. Meanwhile *sdcMicro* has become the standard package for microdata confidentiality and it is widely used by statistical institutions, and for keeping medical data confidential. The advantages of anonymisation of data over the so-called *Remote Execution* are mentioned in this chapter as well as the advantages over the disclosure software μ-Argus [Hundepool et al., 2005]. In the course of the last 18 years μ-Argus has been continuously developed as closed-code by numerous researchers from universities and statistical institutions [Templ, 2008b] who were greatly funded by the European Union and Eurostat (the European statistical institution). It is shown in this book that - in contrast to μ-Argus - not only the application of exploratory method are possible when making data anonymous with *scdMicro*, but also that much more complex methods are implemented making the processing of huge data sets possible.

The package was introduced during several conferences and so far three articles have been published in journals with the most recent [Templ, 2008c] being the Chapter 2 of this book.

Chapter 3 deals with the robustification of disclosure methods. Many SDC-methods for microdata developed so far can be influenced by outliers to a great extent resulting in a high loss of information of the perturbed data. This means that the perturbed data which is made available by researchers turns out to be useless. The chapter describes both the robustification of some methods and some separate methods as well. It was possible to show that only these new methods are useful for the anonymisation of real data. This chapter was published in the Lecture Notes on Computer Sciences Series (Springer, peer-reviewed), see Templ and Meindl [2008b]. The robust methods then developed are included in the R-Packet *sdcMicro*.

Chapter 4 deals with the measurement of the risk of re-identification with numerically scaled variables. It was possible to show that the concepts formulated in literature are useful for multivariate, normally distributed data. The newly developed methods make an applicable handling of outliers, for which a much higher re-identification risk can be assumed, possible. This chapter was published in the Lecture Notes on Computer Sciences Series (Springer, peer-reviewed), see Templ and Meindl [2008a]. The estimation then developed are included in the R-Packet *sdcMicro*.

Chapter 5 again includes several methods of anonymisation and describes a method of making numerically scaled variables anonymous, using robust methods. Furthermore it is proposed to examine loss of information graphically instead of describing it only with numbers.

In the Section "Protection of hierarchical tables" new implementations for keeping up the confidentiality of hierarchical tables are descibed. By creating linear equation systems including marginal conditions, an optimal (minimal) suppression in the tables can be achieved with the help of linear programming - being the solution of these equation systems. Furthermore it was possible to show that an optimal "data attack" for an exact or approximate calculation of cell values can be processed using this implementation. This chapter was published in Templ [2006a]. Some extensions can be found in Templ [2006b]. Most of the methods then developed are included in `sdcTable` which will be freely available in spring 2009.

Chapter 6 contains applications of newly developed methods and software using complex data sets of Statistics Austria. These data sets contain several hundred variables and up to 60 million observations. The data set CVTS2 (*contin-*

uing vocational training survey) is the result of a questionnaire on further training in companies. The anonymisation of this data set is unique because of the enormous amount of categorical variables. Taking this particular structure into account great caution was taken to modify the structure of the data as little as possible and to fulfil the requirements of data privacy. During the anonymisation of the *Austrian tax data* a sample had to be taken a priori and variables had to be selected. Since this data set contained mainly numerically scaled variables, as well as numerous observations, spezialised anonymisations were carried out according to this data set. Researchers can gain access to both anonymous data sets (www.statistik.at/web_de/services/mikrodaten_fuer_forschung_und_lehre). This article will be published in a special edition of Monographs in Official Statistics by Eurostat.

Chapter 7 highlights the application of visualisation tools for the analysis of missing values preceding the choice of an imputation method. Missing values not only play a big part in statistical privacy, they always have to be imputed into official statistics. The quality of data imputation strongly depends on the choice of a suitable imputation method. This choice again depends strongly on the distribution of the missing values in the data. To define such a pattern in the missing values is almost impossible when using ordinary methods. It was possible to show that the newly developed visualisation tool is a useful instrument for recognising special patterns. This chapter was submitted to the *Journal of Statistical Software* and is being revised now. A pre-version was published as a Research Report [Templ and Filzmoser, 2008]. The software developed in connection with it is freely available in the form of an R-Packet with graphical surface [see Templ and Alfons, 2008].

In Chapter 8 new methods for the imputation of composition data are introduced. Due to the linear dependence of the variables from compositional data, distorted estimations are produced when using the Euclidean geometry. New advanced imputation methods are described which do not contain these limitations. It was possible to show that these newly developed methods provide better results than ordinary imputation methods. This was described in a Research Report [Hron et al., 2008b] and submitted to the *Journal of Computational Statistics and Data Analysis*.

Chapter 9 describes the R-Package `robCompositions` [Templ et al., 2009b] which includes both, methods for imputation of missing values in compositional data and exploratory methods for the assessment of the quality of the imputed values. This chapter was printed in the Proceedings-Volume of the conference *New Technologies and Techniques in Statistics (2009)*.

The participation in several European research projects lies at the centre of this book.

Among them the following are listet:

- Centre of Excellence for Statistical Disclosure Control (2006), as project leader for Statistics Austria.

- European Statistical System Network for Statistical Disclosure Control (2008-2009), as project leader for Statistics Austria.

- FP-7 Project AMELI (Advanced Methods for European Laeken Indicators, 2008-2011, project number 217322), as project leader of the Vienna University of Technology (coordinator of the whole project: Ralf Münnich, University of Trier) and as work package leader of two large work packages [see also Templ, 2009].

Zusammenfassung

Der Begriff *Offizielle Statistik* bezieht sich auf jene Fachgebiete der Statistik, welche im wesentlichen in den Statistischen Ämtern praktiziert werden. Die Daten weisen hierbei einige Besonderheiten auf. Stichproben werden von endlichen Populationen gezogen, wobei von diesen endlichen Populationen durchaus Informationen vorhanden sind. Diese Informationen werden in Registern aktualisiert.

Während die Statistischen Ämter ein Monopol in der Erstellung von *Offiziellen Statistiken* aufweisen, werden die Methoden zur Erstellung dieser Statistiken aber nicht nur in den Statistischen Ämtern entwickelt. Viele Forschungsinstitute und Universitäten forschen an verschiedensten Themen in der Offiziellen Statistik: der Methodik zur Überprüfung von Fehlern in den Daten (*Editing*), der Imputation von fehlenden Werten, der Prognose, der Methodik zum Verlinken von Datensätzen mit uneindeutigen Schlüsselvariablen (*Statistical Matching*), der graphischen Aufbereitung von Daten, der Schätzung von Indikatoren, der Methodik zur Schätzung von Statistiken für kleine Gebiete (*Small Area Estimation*) und der statistischen Geheimhaltung von Daten.

Die statistische Geheimhaltung erlangte in den letzten beiden Jahrzehnten große Bedeutung. Die Gesetzeslage, im besonderen das Datenschutzgesetz und das Bundesstatistikgesetz, verlangt, daß bei Weitergabe von Daten an die Öffentlichkeit oder an Forscher keine Rückschlüsse auf statistische Einheiten (z.B. Personen oder Unternehmen) möglich sein dürfen.

In den letzten Jahren wurden von Forschern vermehrt Daten angefragt. Durch diese steigende Nachfrage an *Mikrodaten* (unaggregierte Einzeldatensätze) entwickelte sich in den letzten Jahren das Forschungsgebiet der statistischen Geheimhaltung von Mikrodaten. Ziel ist es, den Datenschutz einzuhalten und dennoch den Forschern brauchbare Daten zur Verfügung stellen zu können. Dies kann durch eine minimale Änderung der Daten erreicht werden, ohne die multivariate Struktur der Daten zu verändern. Die Forschung in der statistischen Geheimhaltung von Mikrodaten wurde speziell in der Statistik Austria im Zuge dieser Arbeit verstärkt [siehe Templ, 2008a].

Statistische Ämter stellen ihre Daten der Öffentlichkeit zumeist in aggregierter Form zur Verfügung. Oft werden die Daten in hierarchischen Tabellen (mit Randsummen) in Publikationen oder auch über Web-Applikationen aufbereitet. Auch hier darf kein Rückschluß auf einzelne statistische Einheiten möglich sein und deshalb müssen bestimmte Werte in Tabellen gesperrt oder verändert werden. Um möglichst wenige Sperrungen oder Veränderungen in den hierarchischen Tabellen durchzuführen und gleichzeitig dem Datenschutz gerecht zu werden, müssen hochkomplexe mathematische Verfahren der linearen Programmierung angewendet werden.

Der richtige Umgang mit fehlenden Werten ist sehr wichtig im Gebiet der statistischen Geheimhaltung, aber auch in allen anderen Fachgebieten der Offiziellen Statistik. Fehlende Werte in der Offiziellen Statistik resultieren aus nichtbeantworteten Fragen eines Fragebogens, aber auch offensichtich falsche Einträge in den Daten werden auf *Missing* gesetzt. Diese fehlenden Werte müssen in Folge geschätzt werden (Datenimputation). Die Güte der Imputation trägt im Wesentlichen zur Qualität der Daten bei und letztendlich bestimmt sie die Qualität jeglicher Indikatoren - wichtig für die Politik und Wirtschaft. Die Auswahl einer (guten) Imputationsmethode hängt von den jeweiligen Daten ab und von der Verteilung der fehlenden Werte. Um in Statistischen Ämtern problembewußtsein zu schaffen und diese wichtigen Problemstellungen aufzuzeigen, ist die Vermittlung des Wissens mit speziellen Kursen (im Speziellen die Verbindung von Computeranwendung und Theorie) wichtig [Dinges and Templ, 2009].

Kapitel 1 enthält eine kleine Einführung in für dieses Buch relevante Methoden, sofern diese nicht in den nachfolgenden Kapiteln erklärt sind. All diese nachfolgenden Kapitel bestehen aus bereits in einschlägigen Fachzeitschriften publizierten oder zur Publikation akzeptierten Artikeln.

Kapitel 2 stellt das entwickelte R-Paket *sdcMicro* [Templ, 2007c, 2008d] vor. Dieses Paket ist frei erhältlich und kann über das *Comprehensive R Archive Network* (CRAN) bezogen werden (http://cran.r-project.org). Mit Hilfe dieses Paketes ist es möglich sehr effektiv Mikrodaten geheimzuhalten. Die Konzepte und die Struktur werden ausführlich erklärt und die Anwendung an Daten gezeigt. Mittlerweile ist *sdcMicro* das Standardpaket zur Geheimhaltung von Mikrodaten und es wird von Statistischen Ämtern, aber auch zur Geheimhaltung von medizinischen Daten, genutzt. Die Vorteile von der Anonymisierung von Daten, gegenüber dem sogenannten *Remote Execution* (Fernrechnen), werden in diesem Kapitel ebenso erwähnt, wie die Vorteile gegenüber der Geheimhaltungs-Software μ-Argus [Hundepool et al., 2005]. μ-Argus wurde in den letzten 18 Jahren durchgehend von zahlreichen Forschern von Universitäten und statistischen Ämtern, mit der großzügigen Subventionen von der Europäischen Union und von Eurostat (dem europäischen statistischen Amt), in Form von *closed-source code* entwickelt [Templ, 2008b]. Es wird gezeigt, daß - im Gegensatz zu μ-Argus - nicht nur eine explorative Vorgehensweise beim Anonymisieren von Daten mit *sdcMicro* möglich ist, sondern auch viel mehr (komplexere) Methoden implementiert sind und das die Verarbeitung von großen Datensätzen möglich ist. Das entwickelte Paket wurde auf mehreren Konferenzen vorgestellt und mittlerweile wurden drei Artikeln in Fachjournalen veröffentlicht - der aktuellste [Templ, 2008c] findet sich im Kapitel 2.

Kapitel 3 beschäftigt sich mit der Robustifizierung von Geheimhaltungsmethoden. Viele der bisher entwickelte Methoden zur statistischen Geheimhaltung von Mikrodaten können durch Ausreißer so stark beeinflußt werden, sodaß dadurch ein hoher Informationsverlust der perturbierten Daten resultieren kann. Das bedeuted, daß die perturbierten Daten, welche den Forschern zur Verfügung gestellt werden, unbrauchbar wären. Das Kapitel beschreibt die Robustifizierung einiger Methoden, aber auch eigenständige robuste Methoden. Es konnte gezeigt werden, daß nur diese neuen Methoden für die Anonymisierung von Echtdaten brauchbar sind. Dieses Kapitel ist der Buchserie *Lecture Notes on Computer Sciences* (Springer, peer-reviewed) veröffentlicht, siehe Templ and Meindl [2008b]. Die entwickelten robusten Methoden sind im R-Paket *sdcMicro* inkludiert.

Kapitel 4 beschäftigt sich mit der Messung des Re-Identifizierungsrisikos bei numerisch skalierten Variablen. Es konnte gezeigt werden, daß die in der Literatur formulierten Konzepte nur für multivariate normalverteilte Daten brauchbar sind. Die neu entwickelten Methoden erlauben den sinnvollen Umgang mit Ausreißern, für welche ein viel höheres Re-Identifizierungsrisiko angenommen werden kann. Dieses Kapitel ist in der Buchserie *Lecture Notes on Computer Sciences* (Springer, peer-reviewed) veröffentlicht, siehe Templ and Meindl [2008a]. Die entwickelten Schätzer sind im R-Paket *sdcMicro* inkludiert.

Kapitel 5 zählt nochmals wiederholend einige Anonymisierungsmethoden auf und beschreibt eine Methode zur Anonymisierung von numerisch skalierten Variablen unter Verwendung von robusten Methoden. Weiters wird erstmals vorgeschlagen, nicht nur Maße für den Informationsverlust zu verwenden, sondern den Informationsverlust graphisch zu beurteilen. Das Kapitel *Protection of hierarchical tables* zeigt neue Implementationen zur Geheimhaltung von hierarchischen Tabellen. Durch das Erzeugen von linearen Gleichungssystemen mit Nebenbedingungen kann mit Hilfe von linearer Programmierung - dem Lösen dieser Gleichungssysteme - eine optimale Sperrung in Tabellen erreicht werden. Ausserdem wird gezeigt, daß mit dieser Implementation sehr leicht ein optimaler "Datenangriff" zur genauen oder näherungsweisen Berechnung von Zellenwerten durchgeführt werden kann. Dieses Kapitel ist in Templ [2006a] erschienen. Einige Erweiterungen sind in Templ [2006b] zu finden. Die entwickelte Methodik ist größtenteils im R-Paket sdcTable inkludiert, welches ab Frühjahr 2009 frei erhältlich sein sollte.

Kapitel 6 beinhaltet eine Anwendung der entwickelten Verfahren und der entwickelten Software anhand von komplexen Datensätzen der Statistik Austria. Die Datensätze inkludieren mehrere hundert Variablen und bis zu 60 Mio. Beobachtungen. Der CVTS2 (*continuing vocational training survey*) Datensatz resultiert aus einem Fragebogen zum Thema Weiterbildung in Unternehmen. Die Anonymisierung dieses Datensatzes ist durch die Fülle von kategorischen Variablen einzigartig. Unter Berücksichtigung dieser Struktur wurde in explorativer Weise versucht, die Struktur der Daten so wenig wie möglich zu verändern und gleichzeitig dem Datenschutz gerecht zu werden. Bei der Anonymisierung der österreichischen Lohnsteuerdaten mußte im vornherein eine Stichprobe gezogen werden und eine Auswahl an Variablen durchgeführt werden. Da dieser Datensatz hauptsächlich numerisch skalierte Variablen, aber viele Beobachtungen inkludiert, erfolgte die Anonymisierung wiederum angepaßt auf diesen Datensatz. Forscher konnen Zugriff auf beide anonymisierte Datensätze erhalten (www.statistik.at/web_de/services/mikrodaten_fuer_forschung_und_lehre). Dieser Artikel erscheint in einem von Eurostat veröffentlichten Sonderband *Monographs in Official Statistics*.

Kapitel 7 unterstreicht die Anwendung von Visualisierungstools zur Analyse von fehlenden Werten vor der Auswahl einer Imputationsmethode. Fehlende Werte spielen nicht nur in der Statistischen Geheimhaltung eine große Rolle. Sie müssen in der Offiziellen Statistik stets imputiert werden. Die Qualität einer Datenimputation hängt stark mit der Auswahl eines geeigneten Imputationsverfahrens ab, wobei diese Auswahl nach der Art der Verteilung der fehlenden Werte in den

Daten abhängt. Eine solche Systematik in den fehlenden Werten festzustellen ist mit herkömmlichen Methoden nur bedingt möglich. Es wird gezeigt, daß die entwickelten Visualisierungswerkzeuge ein gutes Instrument für die Erkennung von speziellen Systematiken ist. Dieses Kapitel wurde in dieser Form im *Journal of Statistical Software* eingereicht und befindet sich in Begutachtung. Eine Vorab-Version wurde als Research Report veröffentlicht [Templ and Filzmoser, 2008]. Die entwickelte Software ist in Form eines R-Paketes mit grafischer Oberfläche auf dem CRAN frei erhältlich [see Templ and Alfons, 2008].

Im Kapitel 8 werden neue Verfahren zur Imputation von Kompositionsdaten vorgestellt. Durch lineare Abhägigkeiten der Variablen von Kompositionsdaten resultieren verzerrte Schätzungen bei der Verwendung der Euklidischen Geometrie. Es werden erstmals neue Methoden zur Imputation beschrieben, welche nicht diesen Einschränkungen unterliegen. Es konnte gezeigt werden, daß diese entwickelten Methoden bessere Ergebnisse liefern als herkömmliche Imputationsmethoden. Dies wurde in einem Research Report beschrieben [Hron et al., 2008b] und im Journal *Journal of Statistical Software* eingereicht.

Kapitel 9 stellt das R-Paket `robCompositions` [Templ et al., 2009b] vor, welches Methoden zur Imputation von fehlenden Werten im Kompositionsdaten sowie explorative Methoden zur Beurteilung der Qualität der imputierten Werte beinhaltet. Dieses Kapitel wurde im Proceedings-Band der Konferenz *New Technologies and Techniques in Statistics (2009)* gedruckt.

Im Zuge dieser Disserationsarbeit stand die Teilnahme an mehreren europäischen Forschungsprojekten im Mittelpunkt. Dabei wurden an folgenden Forschungsprojekten teilgenommen

- Centre of Excellence for Statistical Disclosure Control (2006), als Projektleiter für die Statistik Austria.
- European Statistical System Network for Statistical Disclosure Control (2008-

2009), als Projektleiter für die Statistik Austria.

- FP-7 Projekt AMELI (Advanced Methods for European Laeken Indicators, 2008-2011, Projektnummer 217322), als Projektleiter der TU WIEN (Koordinator des Projektes: Ralf Münnich, Universität Trier) und als *work package leader* zweier zentraler Arbeitspakete [siehe auch Templ, 2009].

Contents

1	**Introduction**		**1**
	1.1	Microdata Protection	3
		1.1.1 Anonymisation of Categorical Variables	4
		1.1.2 Anonymisation of Numerical Variables	14
	1.2	Tabular Data Protection	15
		1.2.1 Primary Sensitive Cells	16
		1.2.2 Secondary Cell Suppression	16
	1.3	Imputation	28
2	**SDC Using sdcMicro**		**31**
	2.1	Introduction	32
		2.1.1 Remote Access	32
		2.1.2 Remote Execution, Model Servers and Automatic Perturbation of Outputs	33
		2.1.3 Data Masking	34
	2.2	Using R for SDC	35
		2.2.1 Reproducibility and Interactivity	35
		2.2.2 Graphical Excellence	36
		2.2.3 Data Formats	36
		2.2.4 Batch Mode and Platform Independence	36
		2.2.5 Explorative Use	37

		2.2.6	Documentation	37
	2.3	Design of Package *sdcMicro*		38
	2.4	Implemented Methods and Data		40
		2.4.1	Methods for Masking Categorical Key Variables	41
		2.4.2	Methods for Masking Numerical Variables	43
		2.4.3	Data Sets Available in the Package	45
		2.4.4	New Methods	45
	2.5	A Small Tour in *sdcMicro*		50
	2.6	Open Source Initiative		59
	2.7	Conclusion		59
3	**Robustification of Masking Methods**			**61**
	3.1	Introduction		62
		3.1.1	Methods Used in this Study	63
		3.1.2	Information Loss	71
		3.1.3	Disclosure Risk	72
	3.2	Robustification of GADP and Shuffling		73
	3.3	Results based on specific artificial data sets		74
	3.4	Simulation		77
		3.4.1	Design of the Simulation Study	77
		3.4.2	Simulation Results	78
	3.5	Conclusion		79
4	**New Disclosure Risk Measures**			**83**
	4.1	Introduction		84
		4.1.1	Distance Based Disclosure Risk Measures	85
	4.2	Special Treatment of Outliers for Disclosure Risk		86
		4.2.1	"Robustification" of *SDID*	87

4.3	New Measures of Disclosure Risk	87
4.4	An Example Using the Tarragona Data Set	92
4.5	Simulation Results	93
4.6	Conclusion	95

5 Software Development for SDC in R — 103

5.1	Using R for Disclosure Control	104
5.2	Microdata protection	106
5.3	Micoraggregation	107
	5.3.1 Clustering at First	108
	5.3.2 Methods Based on Sorting of Variables	109
	5.3.3 Projection Methods and MDAV	109
5.4	Adding Noise	111
	5.4.1 S4-class style	111
	5.4.2 Methods	111
5.5	Other approaches for continuous microdata	112
5.6	Validation of the Results from Microdata Protection	112
5.7	Protection of hierarchical tables	116
5.8	Conclusions	119

6 CVTS2 and Income Tax data — 121

6.1	Introduction	122
	6.1.1 Terms of Use	122
6.2	Software used	123
6.3	CVTS2 Data	123
	6.3.1 Anonymisation of the CVTS2 data	124
6.4	Austrian Income Tax Data	128
	6.4.1 Anonymisation of the Austrian Income Tax Data	129

	6.5	Conclusion	133

7 Visualization of Missing Values — 135

- 7.1 Introduction ... 136
  - 7.1.1 Missing Value Mechanisms ... 139
  - 7.1.2 Limitations for the Detection of the Missing Values Mechanism 140
  - 7.1.3 Detection of the Missing Values Mechanism with Statistical Methods ... 142
- 7.2 Visualization Methods for Missing Values ... 144
  - 7.2.1 Aggregation Plot ... 144
  - 7.2.2 Matrix Plot ... 145
  - 7.2.3 Histogram and Spinogram with Missings ... 147
  - 7.2.4 Scatterplots ... 148
  - 7.2.5 Parallel Coordinate Plots with Missings ... 150
  - 7.2.6 Parallel Boxplots for Missing Values ... 152
  - 7.2.7 Plot Missings in Maps ... 152
- 7.3 R-Package VIM ... 155
- 7.4 Conclusions ... 159
- 7.5 Acknowledgments ... 160

8 Imputation of Missing Values for CoDa — 163

- 8.1 Introduction ... 164
- 8.2 Further properties of compositional data ... 167
- 8.3 Imputation methods for compositional data ... 172
  - 8.3.1 k-nearest neighbor ($k$nn) imputation ... 173
  - 8.3.2 Iterative model-based imputation ... 176
- 8.4 Numerical study with a data example ... 179
- 8.5 Simulation study ... 181

8.6	Conclusions		189
9	**Compositional Data Using the R-Package robCompositions**		**193**
9.1	Introduction		194
	9.1.1	Imputation	194
	9.1.2	Compositional Data	194
9.2	Proposed Imputation Algorithms		196
	9.2.1	k-Nearest Neighbor Imputation	196
	9.2.2	Iterative Model-Based Imputation	196
9.3	Using the R-package robCompositions for Imputing Missing Values		197
	9.3.1	Data	197
	9.3.2	Usage of the Imputation Methods Within the Package	198
9.4	Information Loss, Uncertainty, and Diagnostics		200
	9.4.1	Information Loss Measures	200
	9.4.2	Measuring the Uncertainty of the Imputations	202
	9.4.3	Diagnostic Plots	202
9.5	Conclusions		204

Bibliography   **206**

Index   **236**

CHAPTER 1. INTRODUCTION

1 Introduction to Statistical Disclosure Control and Imputation

Statistical Disclosure Control (SDC) can be divided into three different categories: protection of microdata, protection of tables and remote access / remote execution. The first two are discussed in this introduction in preparation for the following chapters for which these basics are mandatory.

When dealing with data from Official Statistics current laws on data privacy must be considered. In accordance with these laws any disclosure of confidential data to third parties must be avoided. A number of laws regulate these confidentiality aspects for Official Statistics, such as the Austrian legal texts Bundesstatistikgesetz (2000), StF BGBl. I, Nr.163/1999, idF BGBl. I Nr. 136/2001, BGBl. I, Nr. 71/2003, and BGBl. I Nr. 92/2007, or the Commission Regulation 831/2002 and the Commission Regulation 322/1997.

In addition to that, the United Nation Economic Commission report on "Fundamental Principles for Official Statistics" (1992) states: *"Individual data collected by statistical agencies for statistical compilation, whether they refer to natural or legal persons, are to be strictly confidential and used exclusively for statistical purposes"*.

It can be shown that strictly confidentiality implies no information at all. Therefore statistical agencies must loose their obedience towards the statement above. They provide data for which it may only be possible to identify statistical units by

CHAPTER 1. INTRODUCTION

disproportional costs and time resources.

The aim of SDC is to reduce the risk of disclosing information on statistical units (individuals, enterprises, organisations). The aim is to provide as much information as possible (with a minimum of modification (masking) of the data) while respecting confidentiality issues.

Figure 1.1 shows a risk-utility (RU) confidentiality plot [Duncan et al., 2001] which represents the idea of data masking. Masking data means modification of the data in order to avoid disclosure of statistical units. Up to a certain level data which do imply a disclosure risk may be accepted. At the same time the highest possible quality of the data shall be guaranteed, i.e. the data must be protected at a minimum, clever masking of the data. Figure 1.1 shows that the maximum tolerable risk for scientific use files (SUF) is higher than for public use files (PUF) (a more detailed discussion about SUF and PUF can be found in Section 1.1).

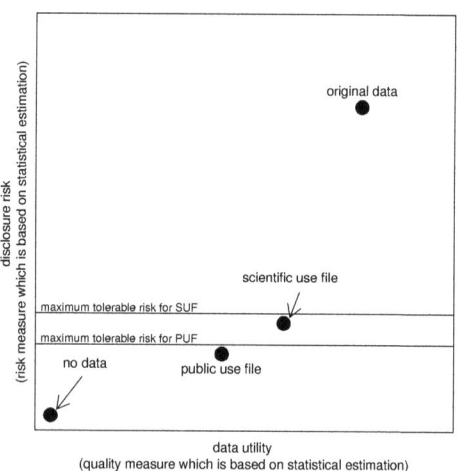

Figure 1.1: Risk-Utility confidentiality map (RU map).

The Relation Between SDC, Imputation and Robustness:
SDC and imputation are closely related, and for both robustness is an important issue. By applying SDC methods some original values are most likely set to be missing for reasons of confidentiality, i.e. some values with high re-identification risk must be set to missing. These missing values are then imputed by the statistical agency or by the data user to ensure that the data can be analysed with standard statistical routines which often need complete data. All possible imputations need to be considered in order to avoid imputations of the same entities as the original ones because such an imputation might then imply a successful disclosure of information.

Robustness of SDC methods and its imputation methods are essential, since with SDC outliers need to be taken special care of. Outliers may be more easily be re-identified as non-outliers [see, e.g., Templ and Meindl, 2008a]. On the other hand, the imputations, however, shall not be driven from outliers but from the main bulk of the data. Chapter 3 and 4 discuss robustness in SDC in more detail.

1.1 Microdata Protection

This is the most recent discipline in SDC. It has proved to be extremely popular and has grown extensively in the last few years. The reason for this is the significant rise in the demand for *public use files* (data which is fully protected) and for *scientific use files* (data which is less protected as public use files) among researchers, institutions and the public.

No global definition of public use files and scientific use files exists. Possible definitions have rather varied over the countries. Generally speaking, public use files are files which are provided to the public, i.e. the data is freely available to everyone. Such data sets must be protected in an extremely restrictive manner so that the disclosure risk of any statistical unit converges to zero. This results in a high information loss. At the same time, the masking of data to obtain scientific use files is less restrictive because the data sets are only provided to researchers who have signed a project-specific contract.

For example, Statistics Austria provides so-called *"standardised data sets"* and *"task related data sets"*, irrespective of the fact that both are scientific use files and both have the same characteristics in the context of SDC.

1.1.1 Anonymisation of Categorical Variables

A first insight into this problem is provided by the following example.

Example 1.1.1[Re-identification of rare combinations] Consider the following situation: assuming that one researcher from the Institute of Statistics and Probability Theory of the Vienna University of Technology has access to microdata from Statistics Austria and that one observation has the following entries in a subset of variables: *residence: Vorau*, *sex: male* and *profession: professor*, whereas *Vorau* is a small village in Styria. Assuming that this researcher happens to know a professor who lives in *Vorau* and that he can now be quite sure that there must be only one professor who is both male and lives in Vorau. The re-identification is successful. After such a re-identification the researcher from the institute is able to read all the entries of this person.

Now it is possible that one additional information which is disclosed after re-identification may be a very sensitive one. The sensitivity of this information depends, of course, on the underlying data which may contain information on diseases or taxes, for example.

Consider a random sample of size n drawn from a population of size N. Let $\pi_i, i = 1, \ldots, N$ be the (first order) inclusion probabilities, i.e. the probability that the element u_i of a population of the size N is chosen in a sample of the size n. Imagine that the re-identification of statistical units could be performed by using external samples or registers with any equal variables, called (categorical) *key variables*.

All possible combinations of categories containing the key variables $X_1, \ldots, X_m$ can be calculated by cross tabulation of these categorical variables. Let $f_k, k = 1, \ldots, n$ be the frequency counts obtained by cross tabulation of the key variables and let F_k be the frequency counts of the population which belong to the same category. If $f_k = 1$ applies the corresponding observation is unique in the sample. If $F_k = 1$ applies then the observation is unique in the population.

A global measure of the re-identification risk is given by the number of sample uniquenesses which are unique in the population as well. It can be expressed in the following way:

$$\tau_1 = \sum_{k=1}^{n} \mathbb{I}(F_k = 1, f_k = 1) \;, \tag{1.1}$$

where $\mathbb{I}$ denotes the indicator function.

Please note, that this notation is different from the common definition of f_k and F_k. The classical notation is based on the formulation of each category, i.e. each cell $k = 1, \ldots, K$ is the cross product of the categories of the key variables, with K being the amount of different categories of the cross product. So, normally, the data is aggregated and the frequency count for each aggregate is estimated. In fact, the two different notations theoretically describe the same phenomenon, but the version used here provides more flexibility which is important for the implementation in software.

When implementing the concept of frequency counts in software this new notation is of great help, since the overall aim is to perform all operations on the whole microdata set and not only on the aggregated data. Applying operations on cells (instead of on each observation) can be rather cumbersome. For example, while plotting the individual risk for each observation, the frequency count for each observation must be reassigned when the traditional concept based on cells instead of observations is used.

Another well-known global risk measure is given by:

$$\tau_2 = \sum_{k=1}^{n} \mathbb{I}(f_k = 1) \frac{1}{F_k} \quad . \tag{1.2}$$

So, the indicator is weighed with the reciprocal value of the population frequency counts. The higher the population frequency count the lower the risk of re-identification. If F_k is particularly high the data intruder cannot be sure if he is able to assign the observation for which he holds information correctly. Hence he is not sure if the re-identification was correct or not.

However, F_k is usually unknown since in statistics only information on samples is collected and therefore little information on the population is known.

Therefore F_k must be estimated in order to be able to estimate τ_1 and/or τ_2, which are then given by

$$\hat{\tau}_1 = \sum_{k=1}^{n} \mathbb{I}(f_k = 1) E(P(F_k = 1 | f_k = 1)) \quad , \quad \hat{\tau}_2 = \sum_{k=1}^{n} \mathbb{I}(f_k = 1) E(\frac{1}{F_k} | f_k = 1) \tag{1.3}$$

F_k may be estimated using the information of the sampling weights which are the inverse of the sample inclusion probabilities. When an observation has a sampling weight equal to 10 it can be assumed that 10 observations do have the same characteristics in the population related to the stratification variables of a (complex) sampling design.

The estimation of the frequency counts in the population is given by the sum of the weights associated with the observations which belong to the corresponding category.

$$\hat{F}_k = \sum_{i \in \{j | x_{j\cdot} = x_{k\cdot}\}} w_i \quad ,$$

with $x_{i\cdot}$ denotes the i-th row of the data consisting of the key variables.

Please note, that this definition is again different from the classical one which expresses F for each category and not directly for each observation.

Example 1.1.2[Frequency counts] In order to demonstrate both the calculation of frequency counts in the sample and the estimation of frequency counts in the population a simple data set is used which is included in the developed version of the R package *sdcMicro* [Templ, 2007c, 2008d].

```
library(sdcMicro)
data(francdat)
x ← francdat[,c(2,4,5,6,8)]
```

The subset x of this data set including the key variables and the sampling weights looks as follows (this data set is also used in Capobianchi et al. [2001]):

	Key1	Key2	Key3	Key4	w_k
1	1	2	5	1	18
2	1	2	1	1	45.5
3	1	2	1	1	39
4	3	3	1	5	17
5	4	3	1	4	541
6	4	3	1	1	8
7	6	2	1	5	5
8	1	2	5	1	92

We now do the calculations and print the output:

```
ff <- freqCalc(x)
cbind(x, ff$fk, ff$Fk)
```

	Key1	Key2	Key3	Key4	w_k	f_k	$\hat{F}_k$
1	1	2	5	1	18	2	110
2	1	2	1	1	45.5	2	84.5
3	1	2	1	1	39	2	84.5
4	3	3	1	5	17	1	17
5	4	3	1	4	541	1	541
6	4	3	1	1	8	1	8
7	6	2	1	5	5	1	5
8	1	2	5	1	92	2	110

It is easy to see that, for example, the values of observation 1 and 8 are equal in the underlying key variables. So, $f_1 = 2$ and $f_8 = 2$. The frequency in the population $\hat{F}_1$ and $\hat{F}_8$ can be estimated with the sum of their sampling weights w_1 and w_8, which equals to 110. Hence, two observations with $x_{k.} = (1, 2, 5, 1)'$ exists in the sample and 110 observations with these entities can be expected in the population.

Unfortunately, it is not reasonable to use $\hat{F}_k$ to estimate the disclosure risk directly. A certain distribution of F_k must be assumed in order to give a realistic measure of individual risk.

The use of prior assumptions for distributions to estimate the individual risk is common in the literature on SDC. Details can be found in Bethlehem et al. [1990], Polletini and Seri [2004], Rinott [1990] and Rinott and Shlomo [2006].

However, all these authors do not provide arguments for F_k being not used directly for the estimation of the individual risk. For better understanding of this problem the following example is given:

Example 1.1.3[Risk estimation] In this example the global risk τ_1 is calculated (from a generated population), but also estimated (from a sample of the population). It shows that F_k should not be used directly to estimate τ_1, for example, and that other concepts for estimating the risk must be considered.

First, a population including four categorical key variables is generated:

```
set.seed(123)
pop <- data.frame(kat1=sample(1:4, 100, replace=TRUE),
                  kat2=round(rnorm(100,6)),
                  kat3=sample(1:3, 100, replace=TRUE),
                  kat4=sample(1:2, 100, replace=TRUE))
```

From this population with 100 observations, a sample with simple random sampling without replacement is drawn. To keep the example as simple as possible no stratification is considered. The inclusion probabilities are fixed to (approximately) $\frac{1}{8}$, and therefore the sampling weights are fixed to 8 due to this experiment.

```
s1 <- round(nrow(pop)/8); sp <- sample(1:nrow(pop), s1)
s <- pop[sp, ]              # the sample
s[,"w"] <- rep(8, nrow(s))  # the corresponding sampling weights
```

In the next step, the frequency counts in the sample are calculated and the frequency counts in the population are estimated. Remember that the frequency counts of the population are exactly known because the population (object pop) is known.

```
popf <- freqCalc(pop, keyVars=1:4, w=NULL)  # frequ. counts of the pop.
samplef <- freqCalc(s, keyVars=1:4, w=5)    # frequ. counts of the sample
```

The parameter w denotes the order of the weight variable in the corresponding data set.

CHAPTER 1. INTRODUCTION

1.1. MICRODATA PROTECTION

The global risk τ_1 can be calculated easily:

```
length(which(popf$Fk[sp] == 1))
R> 6
```

So, the true value is 6, but the estimated risk $\hat{\tau}_1$ is zero:

```
length(which(samplef$fk == 1 & samplef$Fk == 1))
R> 0
```

The previous example has shown that other concepts of estimating risk should be considered. The most popular one is the Benedetti-Franconi Model [Benedetti and Franconi, 1998] which is also implemented in the R-package *sdcMicro*.

The Benedetti-Franconi Model for Risk Estimation

$F_k|f_k$ has to be estimated, i.e. the frequency counts in the population given the frequency counts in the sample. A common assumption is $F_k \sim Poisson(N\pi_k)$ (independently) (see, e.g., Franconi04), where N is assumed to be known and with π_k the inclusion probabilities. (Binomial) Sampling from F_k means that $f_k|F_k \sim Bin(F_k, \pi_k)$. By standard calculations (see, e.g., Bethlehem et al. [1990]) one gets

$$f_k \sim Poisson(N\pi_k) \text{ and } F_k|f_k \sim f_k + Poisson(N(1-\pi_k)) \ .$$

Concerning the risk calculation, the uncertainty on the frequency counts of the population is accounted for in a Bayesian fashion by assuming that the population frequency given the sample frequency, $F_k|f_k$, is negative binomial distributed with success probabilities p_k and number of successes f_k [see also Polletini and Seri, 2004, Rinott and Shlomo, 2006]. By using this assumption (Negative Binomial distribution of $F_k|f_k$), Benedetti and Franconi [1998] estimated the risk τ_2 by the well known and so called "model from Benedetti and Franconi". Using this background Capobianchi et al. [2001] estimate the individual risk $\hat{r}_k$ for each observation as follows

$$\hat{r}_k = \left(\frac{\hat{p}_k}{1-\hat{p}_k}\right)^{f_k} \left\{ A_0 \left(1 + \sum_{j=0}^{f_k-3}(-1)^{j+1} \prod_{l=0}^{j} B_l \right) + (-1)^{f_k} \log(\hat{p}_k) \right\} , \quad (1.4)$$

whereas

$$\hat{p}_k = \frac{f_k}{\hat{F}_k} = \frac{f_k}{\sum_{i \in \{j | x_j = x_k\}} \pi_i} ,$$

while

$$B_l = \frac{(f_k - 1 - l)^2}{(l+1)(f_k - 2 - l)} \frac{\hat{p}_k^{l+2-f_k} - 1}{\hat{p}_k^{l+1-f_k} - 1} \text{ and } A_0 = \frac{\hat{p}_k^{1-f_k} - 1}{f_k - 1} .$$

If $f_k = 1$ Capobianchi et al. [2001] use

$$\hat{r}_k = \frac{\hat{p}_k}{1-\hat{p}_k} \log\left(\frac{1}{\hat{p}_k}\right) ,$$

while if $f_k = 2$ they use

$$\hat{r}_k = \frac{\hat{p}_k}{1-\hat{p}_k} - \left(\frac{\hat{p}_k}{1-\hat{p}_k}\right)^2 \log\left(\frac{1}{\hat{p}_k}\right) ,$$

If the sample is large the computation in formula 1.4 becomes infeasible, but the following approximation works reasonable [Capobianchi et al., 2001]:

$$\hat{r}_k = \frac{\hat{p}_k}{f_k - (1 - \hat{p}_k)}$$

Example 1.1.4[Individual Risk] From our previous example we obtain the following risk by using our developed R-package *sdcMicro*. The individual risk $\hat{r}_k$ is shown in the last column of the following table. The method for the estimation of the individual risk in package *sdcMicro* equals to the concept of Capobianchi et al. [2001] which was outlined in the previous text.

```
xtable(cbind(x, ff$fk, ff$Fk, indivRisk(ff)$rk),
       digits=c(rep(0,4),rep(1,3),3))
```

	Key1	Key2	Key3	Key4	π_k	f_k	$\hat{F}_k$	$\hat{r}_k$
1	1	2	5	1	18.0	2	110.0	0.017
2	1	2	1	1	45.5	2	84.5	0.022
3	1	2	1	1	39.0	2	84.5	0.022
4	3	3	1	5	17.0	1	17.0	0.177
5	4	3	1	4	541.0	1	541.0	0.011
6	4	3	1	1	8.0	1	8.0	0.297
7	6	2	1	5	5.0	1	5.0	0.402
8	1	2	5	1	92.0	2	110.0	0.017

So, the individual risk is large for observation 4, 6 and 7.

Further research has already been done in this topic. Newest approaches consider the neighborhood of each category within log-linear models [see, e.g., Rinott and Shlomo, 2006].

Criticism of the current methods for risk estimation:

Although all the concepts for risk estimation assume a lot of assumptions which may not hold in practice, the most important assumptions which are used by almost all concepts, $F_k \sim Poisson(N\pi_k)$ and $f_k \sim Poisson(N\pi_k)$, may be reasonable in theory but may not be valid in practice.

Further research is needed to formulate more realistic, data-driven measures of risk without assuming prior assumptions about distributions.

Global Recoding and Local Suppression

The table in Example 1.1.4 shows that observation 7 has a high risk of disclosure. This is not surprising because it is unique in the sample and it is estimated that 5 observations with the same entries in the key variables exist in the population.

Therefore, the estimated individual risk must be reduced, e.g., by global recoding of a variable. Here, the categories of certain variables are assigned to broader categories. Doing so, it is very likely that, after the recoding, more observations have the same entries in the key variables and so $\hat{F}_k$ increases and $\hat{r}_k$ decreases.

After such a re-coding it is possible that the risk of re-identification is not reduced or not reduced significantly for several observations. Then, usually, local suppression is applied, where certain values in the key variables are suppressed in order to reduce the risk. This should be done in an optimal way, i.e. to suppress as few values as possible on the one hand and to guarantee low risk of re-identification on the other hand. An iterative algorithm is implemented in the package *sdcMicro* [Templ, 2007c] to search for an optimal solution [Templ, 2007d]. The user can also assign weights to the key variables, because some variables may be less important (such as *age*) as others (such as economic branch classification in business data). Variables with high weights would then be chosen as canditates for possible suppression of some specific values.

Chapter 2 gives a detailed introduction how to apply global recoding and local suppression in an exploratory and interactive manner with the help of the R package *sdcMicro*. In practice, two conditions have to be fulfilled: k-anonymity (with k small) and low individual risk. 4-anonymity, for example, implies that at least 4 observations in the sample have the same entries in the key variables.

Example 1.1.5[Individual Risk (2)] A possible solution of the previous table may look like

	Key1	Key2	Key3	Key4	π_k	f_k	$\hat{F}_k$	$\hat{r}_k$
1	1	2	5	1	18.0	2	110.0	0.017
2	1	2	1	1	45.5	2	84.5	0.022
3	1	2	1	1	39.0	2	84.5	0.022
4	4	3	1	5	17.0	3	30.0	0.048
5	4	3	1	4	541.0	2	549.0	0.003
6	4	3	1	NA	8.0	4	571.0	0.002
7	4	3	1	5	5.0	3	30.0	0.048
8	1	2	5	1	92.0	2	110.0	0.017

Here, the value in the first key variable in row 7 is changed from 6 to 4, and one value is suppressed in the third key variable. As a consequence, the individual risk decreases a lot for certain statistical units.

More examples on real data sets can be found in Chapter 6.

1.1.2 Anonymisation of Numerical Variables

Consider a data set with continuous scaled variables. Then it is clear that almost every observation is unique in the sample considering the numerical variables. Therefore, the concept of uniqueness does no longer work in case of continuous scaled variables.

Unfortunately, a data intruder may have information about a value of a statistical unit. Consider that the intruder knows that the income of one specific person is exactly 47653. If he detects an observation in the data set which has exactly the same value in the variable of interest, then he can be quite sure that the re-identification was successful. The intruder now knows all the information about this person in the data set, i.e. the values of each variable of the sample regarding to this person. Unfortunately, this information could be quite sensitive, for example, information about cancer or taxes. If the data consists of information about enterprises one may even know the turnover of a specific enterprise. It is then possible to re-identify this enterprise if this value occurs in the sample. If the re-identification is successful, the intruder might then know sensitive information about a competitor.

Unfortunately, also when the intruder has only rough information he may know approximately the turnover of a specific enterprise, for example, and a successful match of his data with the sample via *record linkage techniques* (also known as statistical matching) is possible.

Again, the disclosure risk must be determined also for numerical variables using different concepts as for categorical data. Chapter 4 gives a detailed description of this problem, but also new approaches are introduced. To lower the disclosure risk, numerical key variables are usually perturbed. Both, standard perturbation methods and new methods are described in Chapter 2 and Chapter 5.

1.2 Tabular Data Protection

Usually, published tables from statistical agencies are aggregated data, which are represented in tables with the totals in the margins.

A table consists of cells, where each cell consists of an aggregated sum of certain statistical units. Although each cell of a table shows aggregated information of several individuals, the risk of disclosuring statistical units might be high. Table 1.1 illustrates such a disclosure situation. Since only one statistical unit contributes to the cell given by zip-code 5021 and age class $61-65$ (i.e. the cell frequency equals one), the data intruder might be able to know that the salary of this single person is 45801. In addition to that, any of the contributors which corresponds to zip-code 5022 and age class $61-65$ would know the salary of the other. In order to provide a confidential table one has to suppress or change the values of these two cells to avoid a possible re-identification of a statistical unit.

1.2.1 Primary Sensitive Cells

A lot of different rules exist for determining if a cell is (primary) unsafe. The most popular method is the minimum frequency rule. Regarding this rule, a cell is considered safe if more than m statistical units contribute to this cell. Often m is set to 3 or 4.

The (n, k)-rule determines that a cell is unsafe when the total of the n largest contributors exceeds k, whereas k is often fixed to 85%. However, many authors have shown that the p-rule should be preferred (see, e.g., Cox [1981], Willenborg and De Waal [2000] or Merola [2003]). By applying the p-rule, a cell is considered unsafe, when the cell total minus the 2 largest contributors is less than p percent of the largest contribution. These primary suppression rules, but also more sophisticated ones, have been implemented in the unpublished R-package `disclosure` which is used in Chapter 5.

1.2.2 Secondary Cell Suppression

The methodology to determine primary unsafe cells is an easy task in comparison to secondary cell suppression or to other methods for the protection of the primary unsafe cells. Note, that a primary suppressed cell may be re-calculated by considering the linear relationship in the tables.

In practice, we must deal with multidimensional hierarchical and linked tables.

To determine which additional cells must be suppressed (or perturbed) is a complex task and results in difficult combinatorial optimization problems (NP-hard) where a cost function has to be minimized. Such a cost function is expressed as the amount of suppressed cells with the aim to suppress as few cells as possible, but also other cost functions can be chosen. Different heuristics

Table 1.1: Illustration of (primary) sensitive cells in tabular data. (a) shows the median income per age and zip-code. (b) highlights how many individuals correspond to the cells.

	...	56-60	61-65	...		...	56-60	61-65	...
⋮	...	...	...	...	⋮	...	...	...	...
5021	...	44281	45801	...	5021	...	14	1	...
5022	...	37687	38556	...	5022	...	12	2	...
⋮	...	...	...	...	⋮	...	...	...	...

(a) (b)

for 2 or 3-dimensional tables have been proposed (see, e.g., Kelly et al. [1990], Cox [1995], Castro [2002], and Fischetti and Salazar-González [2000]). However, most of the tables from statistical agencies are multidimensional up to 4 to 7 dimensions. A practical implementation with up to four dimensions is implemented in the software τ-Argus. Details on this implementation can be found in Salazar-González [2008a] and Salazar-González [2008b].

However, the computational costs are enormous for large hierarchical tables. De Wolf [2002] uses a heuristic to solve a large amount of smaller linear programs instead of solving the whole linear program at once. This method is often referred as the *hitas* method.

The "Classical" Linear Program

This framework together with the hitas approach is implemented in the R-package `sdcTable` [Meindl, 2009] (former in package `disclosure` [Templ, 2005]) and some results are shown in Chapter 5.

The notation of the linear program is similar to Castro [2002], Fischetti and Salazar-González [2000] and Salazar-González [2008b]. A table can be expressed as an array of cells a_i, $i = 1, \ldots, n$, that satisfies a set of m linear relations

$$My = b \ , \tag{1.5}$$

where M is the matrix of the m linear relations, y corresponds to the cell entries of a marginal table and b are the certain margins of the primary suppressed table. Each y_i fulfils

$$lb_i \leq y_i \leq ub_i \ \forall \ i = 1, \ldots, n \text{ and } \forall i \in \text{PS} \tag{1.6}$$

$$y_i = a_i \ \forall i \notin \text{PS} \ ,$$

with PS the index set of sensitive cells. Usually, lb and ub are pre-defined lower and upper bounds for the sensitive cells.

In the context of SDC, each equation in (1.5) corresponds to a marginal entry, whereas the inequalities in (1.6) are closed bounds which an attacker can know by attacking a table. Matrix $M \mid m_{ij} \in \{-1, 0, 1\}$ has dimension $(n \times m)$, with n representing the cells and m representing the equations.

Example 1.2.1[Attacking a Table] To illustrate the generation of M a simple problem is considered - the attacker problem. Suppose that an intruder wants to calculate the entries of a suppressed table. The optimal way to solve this problem is to apply a linear program, i.e. to minimize and maximize an unknown cell under the constraints given by M.

	A	B	C	total
I	104	P (47)	S (15)	166
II	21	P (3)	P (2)	26
III	302	122	35	459
IV	427	172	52	651

In this table four cells are suppressed, whereas P is denoted for primary suppressed cells and S for secondary suppressed cells. The original (unsuppressed) values are highlighted in brackets.

Let Y denote the entries of this table and y_1 the cell given by category A and activity I, y_2 the cell given by category B and activity I, ..., y_4 the cell given by category A and activity II, ..., and y_9 the cell given by category C and activity III.

Matrix M can be chosen as follows (using our developed package *disclosure*):

```
library(disclosure)
attack <- lp2.hier(tab)$m
attack$linear
```

	1	2	3	4	5	6	7	8	9
1	0	1	1	0	0	0	0	0	0
2	0	0	0	0	1	1	0	0	0
3	0	0	0	0	0	0	0	0	0
4	0	0	0	0	0	0	0	0	0
5	0	1	0	0	1	0	0	0	0
6	0	0	1	0	0	1	0	0	0

The first line of matrix M highlights that in the first row of the table y_2 and y_3 are missing and all other y's are not missing with $y_i, i \in \{1,\ldots,9\}$ consisting of the cells without the margins. The second line of M highlights that in the second row of the previous table, y_5 and y_6 are missing. The fourth to the sixth row corresponds to the columns of the table. For example, the fifth row of matrix M highlights that y_2 and y_6 are missing in the second column of the underlying table.

The right hand side of the equation (1.5), b, is given by the margins of the table:

```
attack$l.rhs
62  5  0  0 50 17
```

So, the value of 62 is obtained by $166 - 104$, 5 is calculated by $26 - 21$, 0 by $459 - 302 - 102 - 35$, the next 0 by $427 - 104 - 21 - 302$, etc.. To determine the upper bound and lower bound of y_2 which can be known by the attacker, for example, one has to minimize and maximize y_2, i.e.

$$\min y_2 \text{ and } \max y_2$$

given the constraints

$$My = b$$

An attack using the developed routines in R gives the following bounds by solving this linear program:

```
attack$lp.out2
```

	min	max	true	nrow	ncol	ind
1	45	50	47	1	2	1
2	12	17	15	1	3	0
3	0	5	3	2	2	0
4	0	5	2	2	3	0

Here, the primary suppressed cell given by category B and activity I is approximately known by the attacker, i.e. the attacker knows that the true value must lay in the interval $[45, 50]$ which is too close in practice. Therefore additional cells have to be suppressed to broaden this interval.

CHAPTER 1. INTRODUCTION
1.2. TABULAR DATA PROTECTION

The Integer Linear Program

Previously, the solution of the attacker's problem was shown, where the attacker simply can obtain upper and lower bounds for a sensitive cell. To formulate the most simple linear program for solving the secondary cell suppression problem a simple numerical table from Salazar-González [2005] is considered. The table in Example 1.2.2 highlights the cell values of this table and its notation.

Example 1.2.2[Simple Table and its Notation]

W	A	B	C	Total
I	20	50	10	80
II	8	19	22	49
III	17	32	12	61
Total	45	101	44	190

W	A	B	C	Total
I	y_1	y_2	y_3	y_4
II	y_5	y_6	y_7	y_8
III	y_9	y_{10}	y_{11}	y_{12}
Total	y_{13}	y_{14}	y_{15}	y_{16}

Please note, that y_i is denoted in a different way as for the attacker problem described in Example 1.2.1, because this problem is much harder to solve and different strategies must be applied.

From the corresponding Table the following conditions can be easily obtained:

CHAPTER 1. INTRODUCTION

1.2. TABULAR DATA PROTECTION

$$\mathbf{y_4} = y_1(20) + y_2(50) + y_3(10) = 80$$
$$\mathbf{y_8} = y_5(8) + y_6(19) + y_7(22) = 49$$
$$\mathbf{y_{12}} = y_9(17) + y_{10}(32) + y_{11}(12) = 61$$
$$\mathbf{y_{13}} = y_1(20) + y_5(8) + y_9(17) = 45$$
$$\mathbf{y_{14}} = y_2(50) + y_6(19) + y_{10}(32) = 101$$
$$\mathbf{y_{15}} = y_3(10) + y_7(22) + y_{11}(12) = 44$$
$$\mathbf{y_{16}} = y_4(80) + y_8(49) + y_{12}(61) = 190$$
$$\mathbf{y_{16}} = y_{13}(45) + y_{14}(101) + y_{15}(44) = 190$$

A (multi-dimensional, hierarchical) table is given by

- a data vector $a = [a_1, \ldots, a_n]$
- constraints which can be expressed in matrix notation: $Ma = b$
- upper and lower bounds for each cell: $lb_i \leq a_i \leq ub_i$
- p primary suppressed cells: $PS = \{i_1, \ldots, i_p\}$

Example 1.2.3[Matrix M] Within the previous example, the matrix M is given by $M_{ij} \in \{-1, 0, 1\}$ and $b = 0$:

$$M = \begin{pmatrix} 1 & 1 & 1 & -1 & 0 & 0 & 0 & 0 & 0 & 0 & 0 & 0 & 0 & 0 & 0 \\ 0 & 0 & 0 & 0 & 1 & 1 & 1 & -1 & 0 & 0 & 0 & 0 & 0 & 0 & 0 \\ 0 & 0 & 0 & 0 & 0 & 0 & 0 & 1 & 1 & 1 & -1 & 0 & 0 & 0 & 0 \\ 0 & 0 & 0 & 0 & 0 & 0 & 0 & 0 & 0 & 0 & 0 & 1 & 1 & 1 & -1 \\ 1 & 0 & 0 & 0 & 1 & 0 & 0 & 0 & 1 & 0 & 0 & -1 & 0 & 0 & 0 \\ 0 & 1 & 0 & 0 & 0 & 1 & 0 & 0 & 0 & 1 & 0 & 0 & -1 & 0 & 0 \\ 0 & 0 & 1 & 0 & 0 & 0 & 1 & 0 & 0 & 0 & 1 & 0 & 0 & -1 & 0 \\ 0 & 0 & 0 & 1 & 0 & 0 & 0 & 1 & 0 & 0 & 0 & 1 & 0 & 0 & -1 \end{pmatrix}$$

Consider that in table from Example 1.2.2 y_7 is primary protected. Here, the primary suppressed cell value can be calculated very easily by substracting the total with certain other cell values, namely $49 - 8 - 19 = 22$. Because of the linear relationships in the table it is necessary to suppress other cell values as well to protect this primary protected cell. Whereas it is easy to find such secondary suppressions for the underlying table, the problem is not solvable in reasonable time (it is NP-hard) for larger, hierarchical and multi-dimensional tables. Optimal solutions for the secondary cell suppression problem are based on linear programs, whereas optimality implies that both, a minimum amount of cells is suppressed (or by minimizing another cost function) and the table is safe, must be fulfilled. A primary suppressed cell can be considered as safe, if this cell can be estimated only within a pre-defined interval (e.g. $\pm 10\%$ around the original cell value).

Before the integer linear problem can be formulated, the problematic "and" condition in Equation (1.6) must be re-formulated.

Consider that for each cell value a_i an attacker can obtain a lower and a upper bound (lb_i and ub_i), whereas

$$lb_i \leq a_i \leq ub_i \quad \forall i = 1, \ldots, n \; .$$

Then, for each cell the lower and upper bounds are defined as follows:

$$LB_i := a_i - lb_i \geq 0 \quad \forall i = 1, \ldots, n$$
$$UB_i := ub_i - a_i \geq 0 \quad \forall i = 1, \ldots, n$$

For each sensitive cell, lower (LPL_I) and upper (UPL_i) protection levels are defined in such a way that the attacker's estimated bounds fulfill

$$min(y_i) \leq a_i - LPL_i \quad \forall i \in PS$$
$$max(y_i) \geq a_i + UPL_i \quad \forall i \in PS.$$

A binary variable x_i, $i = 1, \ldots, n$ is defined, which fulfils:

$$x_i = 0 \; \forall i \notin SUP$$
$$x_i = 1 \; \forall i \in SUP,$$

for a given *suppression sample* (SUP) which is an index set denoting the (currently) suppressed cells, i.e. the primary suppressed cell and the secondary suppressed cells.

We also have to define a weight w_i for each cell a_i which influences the an objective function (denoted in the following). Each of the following weights are reasonable

$$w_i = a_i$$
$$w_i = 1$$
$$w_i = log(1 + a_i)$$

Then the objective function for the optimization problem is given by

$$min \sum_{i=1}^{n} w_i \cdot x_i \quad , \qquad (1.7)$$

and the corresponding contraints are given by

$$Mf = b \qquad Mg = b \qquad (1.8)$$
$$f_i \geq a_i - LB_i \cdot x_i \; \forall i = 1, \ldots, n \quad g_i \geq a_i - LB_i \cdot x_i \; \forall i = 1, \ldots, n \qquad (1.9)$$
$$f_i \leq a_i + UB_i \cdot x_i \; \forall i = 1, \ldots, n \quad g_i \leq a_i + UB_i \cdot x_i \; \forall i = 1, \ldots, n \qquad (1.10)$$
$$f_i \leq a_i - LPL_i \; \forall i \in PS \qquad g_i \geq a_i + UPL_i \; \forall i \in PS \quad . \qquad (1.11)$$

In the following, two tables $f = (f_1, \ldots, f_n)$ and $g = (g_1, \ldots, g_n)$ must be calculated. The constraints (1.8), (1.9), (1.10) are needed to ensure that all linear dependencies are considered, and that:

$$f_i = g_i = a_i \; \forall i \notin SUP \qquad (1.12)$$
$$lb_i \leq f_i, g_i \leq ub_i \; \forall i \in SUP \quad . \qquad (1.13)$$

CHAPTER 1. INTRODUCTION
1.2. TABULAR DATA PROTECTION

Concerning that problem, those cells are secondary suppressed so that both conditions hold, that an attacker can only find an interval for a primary suppressed cell for which he only know that the true value must be inside this interval and that the objective function is minimized. The obtained suppression sample SUP indicates the indices of all the suppressions which fulfills these conditions.

The constraints given in (1.11) enforce to keep the required protection levels for all primary suppressed cells.

However, in reality, the amount of variables and constraints for a huge hierarchical table is often to large to be able to obtain an optimal solution, i.e. the corresponding matrix M is larger than 50000×50000 which makes it impossible to solve it in reasonable time. However, it is possible to reduce the dimensionality of M by formulating the dual problem of the integer linear problem given in (1.7)-(1.11). Also heuristic procedures can be formulated which do not solve the problem optimal, but where an adequate near optimal solution is obtained.

Example 1.2.4[Hierarchical Structure of a Table] To give an impression how complex the problem with hierarchical tables could be, a very simple - only 2-dimensional - small table is chosen:

	A	B	C	Total
6211	20	50	10	80
6212	8	19	22	49
6213	17	32	12	61
T621	45	101	44	190
62221	9	28	5	42
62222	4	7	6	17
62223	27	15	9	51
6222	40	50	20	110
6223	2	20	18	40
6224	20	30	25	75
T622	62	100	53	225
T62	107	201	97	415

So, the table splits up into some (business) classification codes (whereas 6211, 6212 and 6213 sum up to T621, for example) and activities (A,B and C). Note, that

CHAPTER 1. INTRODUCTION

1.2. TABULAR DATA PROTECTION

T621 and T622 sum up to T62.

Example 1.2.5[Protection of a Hierarchical Table] Consider the previous table and that the following cells must be primary protected.

	A	B	C	Total
6211	20	50	10	80
6212	8	19	NA	49
6213	17	32	12	61
T621	45	101	44	190
62221	9	28	5	42
62222	NA	7	6	NA
62223	27	15	9	51
6222	40	NA	20	110
6223	NA	20	18	40
6224	20	30	25	75
T622	62	100	53	225
T62	107	201	97	415

Afterwards one has to look very closely to find a minimum amount of secondary suppressions in order to protect the primary cells. One possible solution could be:

	A	B	C	Total
6211	20	50	10	80
6212	S	19	NA	49
6213	S	32	S	61
T621	45	101	44	190
62221	9	28	5	42
62222	NA	S	6	NA
62223	27	15	9	51
6222	S	NA	20	S
6223	NA	S	18	S
6224	20	30	25	75
T622	62	100	53	225
T62	107	201	97	415

By additionally suppressing these 8 cells, the sum of these suppressed cell values is 254. Often different solutions for suppressing a table can be found with the same amount of secondary suppressed cells. It is obvious that a solution with 8 additional suppressed cells for which the sum of the suppressed cells is lower than 254 should be prefered, because then, in general, more information is provided.

It is easy to find a better solution, such as:

	A	B	C	Total
6211	20	50	10	80
6212	S	19	NA	49
6213	S	32	S	61
T621	45	101	44	190
62221	S	28	5	S
62222	NA	S	6	NA
62223	27	15	9	51
6222	S	NA	20	110
6223	NA	S	18	40
6224	20	30	25	75
T622	62	100	53	225
T62	107	201	97	415

Here, we also suppress additionally 8 cells, but the sum of the additional suppressed cell values is only 155.

By solving a linear program, such as given by the objective function in (1.7) and the corresponding constraints given in (1.8), (1.9), (1.10), an optimal solution for the previous example with respect that each primary suppressed cell is obtained. An attacker can then only determine the cell values of the primary suppressed cells within a certain interval.

In practice, the tables are often up to 7-dimensional with complex hierarchies which makes it impossible to find any suppression without sophisticated, computationally expensive methods.

Similar problems can be formulated for other approaches, such as *Controlled*

Tabular Adjustment (see, e.g., Castro and Baena [2008]).

1.3 Imputation

The presence of missing values in data from Official Statistics is quite common. In virtually every survey sample item non-responses (if one question is not answered by a responded, for example) or unit non-responses (a whole battery of questions is not answered by a respondent) occurs. Another example for missing values has been given in the previous section, where values were set to missing for data protection. Consequently, statistical agencies often provide estimates of the missing values together with an information if a value is imputed or not. This helps the data user who often do not want to impute the missing values because frequently it is not easy to apply a proper imputation. Therefore, this topic is highly related with the previous section, because often both, anonymisation and imputation, must be applied before releasing the data. But also possible imputations made by data users should be considered during the application of SDC methods.

Most statistical methods cannot be directly applied to data sets with missing observations. While in the univariate case the observations with missing information could simply be deleted, this can result in a severe loss of information in the multivariate case. Multivariate observations usually form the rows of a data matrix, and deleting an entire row implies that cells carrying available information are lost for the analysis. In both cases (univariate and multivariate), the problem remains that valid inferences can only be made if the missing data are *missing completely at random* (MCAR) (see, e.g., Little and Rubin [1987]). Instead of deleting observations with missing values it is thus better to fill in the missing cells with appropriate values. This is only possible if additional information is available, i.e. in the multivariate case. The estimation of missing values is also known under the name *imputation* (see, e.g., Little and Rubin [1987]). Once all missing values have been imputed, the data set can be analyzed using the standard techniques for complete data.

Many different methods for imputation have been developed over the last few decades. While univariate methods replace the missing values by the coordinate-wise mean or median, the more advisable multivariate methods are based on similarities among the objects and/or variables. A typical distance based method is k-nearest-neighbor (kNN) imputation, where the information of the nearest $k \geq 1$ complete observations is used to estimate the missing values. Another well-known procedure is the EM (expectation maximization) algorithm Dempster et al. [1977], which uses the relations between observations and variables for estimating the missing cells in a data matrix. Further details, as well as methods based on multiple regression and principal component analysis are described in Little and Rubin [1987] and Schafer [1997].

Most of these methods use assumptions, such as *missing at random* (MAR) (see, e.g., Little and Rubin [1987]). Moreover, one usually assumes that the data originate from a multivariate normal distribution, which is no longer valid in presence of outliers in the data. In this case the "classical" methods can give very biased estimates for the missing values, and it is more advisable to use robust methods, being less influenced by outlying observations (see, e.g., Beguin and Hulliger [2008], Serneels and Verdonck [2008], Hron et al. [2008a], Templ et al. [2009a]).

For estimating the missing values in an optimal way, visualization of data with missing values should be carried out. Here, with the help of proper visualization tools the data analyst can visualize the missing values itself in order to detect the missing values mechanism for each variable in each category. After this pre-study, proper imputation strategies can be chosen. In Chapter 7 it is explained why it is recommended to use such visualization tools and how it helps for imputation. In this chapter also new methods for the visualization of missing values are introduced. For an effective application of the methods a a graphical user interface is provided [Templ and Alfons, 2008].

2 Statistical Disclosure Control for Microdata Using the R-Package sdcMicro

Published in "Transactions on Data Privacy" [Templ, 2008c]

Matthias Templ[*,**]

[*] Department of Methodology, Statistics Austria, Guglgasse 13, 1110 Vienna, Austria. (matthias.templ@statistik.gv.at) and

[**] Department of Statistics and Probability Theory, Vienna University of Technology, Wiedner Hauptstr. 8-10, 1040 Vienna, Austria. (templ@statistik.tuwien.ac.at)

Abstract: The demand for high quality microdata for analytical purposes has grown rapidly among researchers and the public over the last few years. In order to respect existing laws on data privacy and to be able to provide microdata to researchers and the public, statistical institutes, agencies and other institutions may provide masked data. Using our flexible software tools with which one can apply protection methods in an exploratory manner, it is possible to generate high quality confidential (micro-)data.

In this paper we present highly flexible and easy to use software for the generation of anonymized microdata and give insights into the implementation and the design of the **R**-Package *sdcMicro*. **R** is a highly extendable system for statistical computing and graphics, distributed over the net. *sdcMicro* contains almost all popular methods for the anonymization of both categorical and continuous variables. Furthermore, several new methods have been implemented. The package can also be used for the comparison of methods and for measuring the information loss and disclosure risk of the masked data.

Keywords: Statistical Disclosure Control, Microdata, Software Development, R.

2.1 Introduction

Nowadays a number of different concepts exist to make confidential data accessible to researchers and users. A short outline of these concepts is given below in order to point out the need for data masking and the need for flexible software for data masking.

2.1.1 Remote Access

The most flexible way to provide microdata may be by installing *remote access* facilities where researchers can have a look at the data using a secure connection. They can analyse the data and choose a suitable model, but it is not possible for them to download it. Implementations can be found in Hundepool and de Wolf [2005] or Borchsenius [2005]. Unfortunately, in some countries remote access can only be partially applied, depending on the discipline from which the data originates. This is due to different legacy for data coming from different disciplines. In Austria, for example, only microdata masking or *remote execution* is applicable for official statistics because of the legal situation which prohibits viewing "original" data.

2.1.2 Remote Execution, Model Servers and Automatic Perturbation of Outputs

When remote execution is applied, it is often necessary to generate synthetic data. Researchers can build and try out their methods using these synthetic data. Only in the final stage these methods are applied to the original data by the data holders. At this stage the output must be checked to detect and to avoid confidentiality disclosure. Without such synthetic data the data holders would have to check the output at every request which is very time-consuming and expensive. In addition to that, output checking is a highly sophisticated process in which the data holders may never be able to be sure f certain results lead to disclosure or not. Furthermore, data holders might never be able to understand all methods (for example available in more than 1500 **R**-packages) which might be applied to the data. Therefore, output checking may not be feasible at all.

Something between remote access and remote execution are the so-called *confidential preserving model servers* (Steel and Reznek [2005]). Users are able to apply models to data which cannot be seen. Graphical summaries and parameter estimates of the data or of the resulting models are often perturbed (see e.g. in Heitzig [2006]) and especially outliers must be treated in a special way. Outlier detection itself is essential in many practical applications and may hardly be supported within this approach. Data almost always include outliers which can only be detected by robust procedures, i.e. to apply the whole range of diagnostic tools on robust estimates. Naturally, such robust estimates cannot be applied since the re-identification of outlying observations may be successful and the perturbation of outliers will make it impossible to select a model and a suitable method. On the other hand, the perturbation of summaries of non-smooth statistics - the approach of Heitzig [2005] - leads to serious problems. More details on this problem can be found in Templ [2007a]. Another drawback is that only few methods can be implemented in such model servers within a reasonable time because every method must be adapted to avoid disclosure results when the users are applying these methods.

2.1.3 Data Masking

Our aim is to provide access to masked data since remote access cannot be applied due to legal restrictions in many cases and remote execution is too cost-intensive and time-consuming. We want to provide data which can be analysed by the researchers with their own methods and using their own software.

A popular software package, which has implemented some methods for masking data, is μ-Argus Hundepool et al. [2005]. This software provides a graphical user interface (GUI) which may help users with little experience with statistical software to produce safe microdata sets.

In this paper we show that data masking can be easily performed with the newly-devel- oped and flexible software package *sdcMicro* by minimizing information loss and the risk of re-identification. This software can be downloaded from the comprehensive **R** archive network on *http://cran.r-project.org* (Templ [2007d]).

In addition to that, we also concentrate on the design of this new package for microdata protection and illustrate the open source philosophy of this project.

The outline of this contribution is as follows:
In chapter 3 we motivate why we have implemented methods for SDC in **R**. Chapter 4 briefly describes the methods which we have implemented in our package. We distinguish between methods for categorical and continuous scaled variables, followed by specific design issues of *sdcMicro*. In chapter 5 we show a practical application on real data in order to demonstrate the usage of the package. In chapter 6 we describe the open-source philosophy of our package and the bene-

fits of such an open-source project related to SDC. Finally, we give a conclusion pointing out the capacity of package *sdcMicro*.

2.2 Using R for SDC

R R Development Core Team [2008a] is an open source high-level statistical computing environment subjected to the General Public License and therefore freely available and extendable. Furthermore, **R** has become the standard statistical software and thousands of people are involved in the development of **R** both at universities and companies. More than 1400 add-on packages have been created over the last years.

We will now discuss and present the usefulness of **R** with respect to SDC.

2.2.1 Reproducibility and Interactivity

The most effective way of masking microdata is by doing the anonymization steps in an exploratory and in some sense iterative way. It is possible to apply various methods on various variables with different parameters resulting in different effects on the data while looking for sufficient anonymization of the data with respect to low information loss and low re-identification risk. Therefore, we must be able to reproduce any step of the anonymization process easily. This is possible by running a script which includes all the necessary commands. It is then easy to change and adapt the script, to run it and to get the updated results "in real time". Of course, this is fulfilled by many software tools but the real advantage of using **R** is that we can interactively "play" with the data, i.e. access all the objects in the workspace of **R** at any time and we can change, display or apply operations interactively on these objects on the fly. This is very useful during the anonymization process and is a quite different concept than writing a "batch file" which can then be executed.

2.2.2 Graphical Excellence

Powerful easy-to-grasp visualization tools display the effects of the anonymization on the data not only in order to compare the masked with the original data. For an effective exploratory approach during data anonymization such visualization tools are very helpful during the anonymization procedure because the effects of methods or parameter adjustments can be seen interactively. It is often more informative to look at some (comparison) plots to get a feeling about the information loss when perturbing the data, and to explore graphically distributions and the multivariate structure of the masked data and the original data instead of computing certain measures of information loss (see e.g. in Templ [2006a] and the *scdMicro* package manual given in Templ [2007d]).

2.2.3 Data Formats

Since numerous different data formats are handled by users it is very convenient to have the opportunity to import and export data formats from data base software like `MySQL`, `ODBC` database access, etc., statistical software like `SPSS`, `SAS`, `Stata`, "`Excel`" and data from other formats like `ASCII`, fixed format files and many more, supported within the **R** data import and export facilities (for more information see R Development Core Team [2008b]).

2.2.4 Batch Mode and Platform Independence

Sometimes it is also useful to run the anonymization tools via batch mode which can be easily derived with the package *sdcMicro* Templ [2007d]. It is also very important to provide software for SDC which is platform independent and works on all common operating systems.

2.2.5 Explorative Use

When perturbing categorical key variables - the variables on which an intruder may (partially) know and which may be used for re-identification by cross tabulation - recoding and local suppression come to mind. On the one hand the SDC-specialists try to recode as few as possible variables, and on the other hand as few as possible local suppressions should be made. Naturally, it is a highly exploratory process to find a way to recode only a subset of variables in a specific way in order to require only few suppressions, and at the same time to guarantee low individual risk and/or k-anonymity (Sweeney [2002]) [1]. As mentioned before, **R** is very well suited for interactive playing with the data and for trying out many different opportunities in "real time".

When perturbing numerical data the optimal perturbation method strongly depends on the multivariate structure of the data. Therefore, a flexible tool with which one can try out various methods and easily compare the methods is necessary. Additional flexibility may be provided by giving the user the opportunity to include his own functions for SDC.

2.2.6 Documentation

The provided **R** functions may be self-exploratory because the interested user can easily understand the method by looking at the source code. However, additional documentation is provided in a standardized way which includes the description of the method (or provides references, in case a detailed description is out of reach), the description of the function arguments, the description of the output as well as self-explaining executable examples. Furthermore, more general documentation with practical examples is available as a package vignette explaining parts of the

---

[1] If the frequency count of every observation is greater than k. The frequency count for an observation is defined as the number of observations featuring the same values with respect to the key variables.

package in a more informal way than the strict format of the usual **R** help files does by showing possible interactions of functions in the package with the help of practical examples.

All these topics are supported when using the software system **R**. In addition to that, dynamical reports can be used for documenting the anonymization process by using **R** in combination with Sweave (Leisch [2002a]). Hereby, the **R**-code for anonymization is directly placed in LaTeX files which are compiled using Sweave. Reports can be generated very effectively and quickly for particular steps in the production process, or to provide some fully-documented reports including graphics and **R** output for different sets of parameters. Using a fully-developed, object-oriented, open-source programming language is a great advantage for many reasons which are outlined in the following chapter. In addition to that, very large data sets can be processed in low computation time with this package.

2.3 Objectives in the Design of Package *sdcMicro*

The advantages of using an object-oriented programming language become apparent in the *scdMicro* package. However, **R** is not a pure class-oriented programming language. It can be seen as a function- and class-oriented programming language which allows for a greater programming flexibility.

In **R** everything is an object and every object belongs to a specific class. The class of an object determines how it will be treated, and generic functions perform either a task or an action on its arguments according to the class of the argument itself (see e.g. in R Development Core Team [2008c]). The class mechanism offers a facility for designing and writing functions for special purposes to the programmers which is extensively used in *sdcMicro*. Nearly all functions, e.g. the one for the individual risk methodology or the frequency calculation, produce objects from certain classes. Different *print*, *summary* and *plot* methods are provided for each of these objects depending on their class. plot(ir1) produces a completely different result than plot(fc1) assuming that the objects ir1 and fc1 are objects from different classes, i.e. resulting from different functions in package *sdcMicro*. This object-oriented approach allows simple use of the package to any user, independently of the proficiency in **R**. Furthermore, users can try out methods with several different parameters and they can easily compare the methods with the implemented summary and plot methods.

Note that no metadata management needs to be done by the user, even not after importing the data into **R**. You can apply the methods directly on your data sets or on objects from certain classes. The only thing that has to be done is to determine which of the variables should be considered as the key variables, weight vector and as the confidential numerical variables.

An online documentation is included in the package containing all explanations on all input and output parameters of every function. Furthermore, various examples are included for each of the functions. All the examples can be easily executed by the users. Since *sdcMicro* is checked automatically every day from the CRAN server it is guaranteed that these examples work. Furthermore, additional checks are executed. E.g. the consistency of the documentation and the functions can be guaranteed.

To be able to deal with large data sets, expensive computational calculation steps are implemented in C and included in **R** via the R/C++ interface. The estimation of the frequency counts for the population is one of the most critical calculation steps regarding the computation time. For example, Figure (2.1) shows the calculation time for frequency counts using the relatively small μ-Argus test data set with only 4000 observations (available within the anonymization software μ-Argus, see Hundepool et al. [2006], version 4.1.0). It is carried out by a three-year old personal computer[2] running the Windows XP operating system. In order to estimate population frequency counts it is necessary to program with nested loops which is computationally expensive. It is therefore not feasable to run these estimations on large data sets using traditional for-loops in **R** because of possible long running times and required memory. Even when trying to avoid such for-loops in **R** using advanced concepts for data manipulation (using function apply, for example), the running time and the required memory is still a problem. Only when using advanced concepts of data manipulation in **R** for which the underlying functions are written in C or Fortran it is possible to calculate the frequency counts with up to 6 key variables in an adequate time. This is not possible for the μ-Argus system which runs out of memory with 5 or more key variables. The developed R/C can also handle a enormous number of key variables[3] in a reasonable time, even when doing the calculation on much larger data sets than the μ-Argus test data set. In Figure (2.1) you can easily see that the computation time is always less than 1 second for this small data set, and it remains low for much larger data sets.

2.4 Implemented Methods and Data

Figure 2.2 shows the main structure of package *sdcMicro*. After reading the data (from a certain format) into **R** the user is able to protect the categorical key variables and/or to protect the confidential numerical variables. In addition to that, several functions are provided for the generation of synthetic data and for the measurement of the disclosure risk and the data utility for numerical data.

---

[2]Intel x86 based system with 3Ghz and 1 GB memory
[3]to choose such a large number of key variables is, of course, not always meaningful.

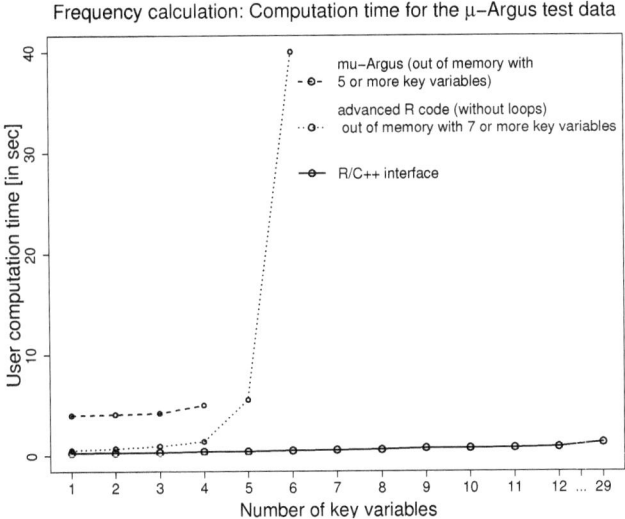

Figure 2.1: Computation time for calculating the frequency counts for the μ-Argus test data set with 4000 observations.

2.4.1 Methods for Masking Categorical Key Variables

Function freqCalc() calculates the frequency count for each observation as well as the frequency counts with respect to the sampling weights (details and a illustrative example can be found in Franconi and Polettini [2004a]). Function freq-Calc() returns an object of class *freqcalc*. For objects of this class a print and a summary method is provided (see Figure 2.2). Objects of class *freqCalc* can be used as input parameters to the functions globalRecode() and indivRisk(). globalRecode() provides some functionality for recoding variables. indivRisk() estimates the individual risk for re-identification as implemented in μ-Argus (see e.g. Franconi and Polettini [2004a]) and produces objects of class *indivRisk* for which a plot and a print methods are implemented. Concerning the risk calcula-

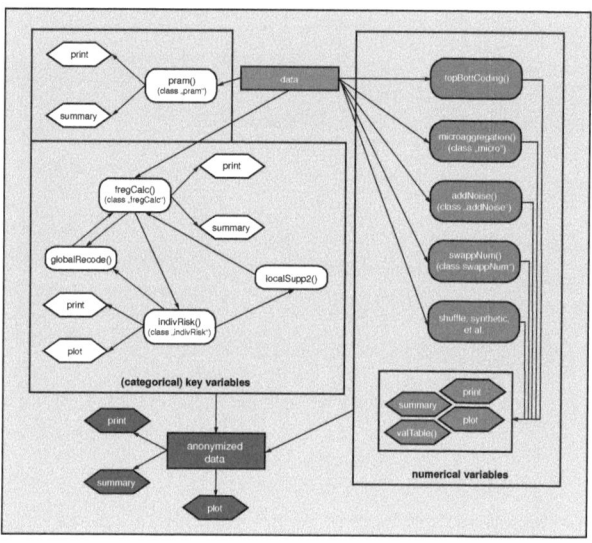

Figure 2.2: Survey on certain procedures in package *sdcMicro* and their relationship.

tion, the uncertainty on the frequency counts of the population is accounted for in a Bayesian fashion by assuming that the population frequency given the sample frequency is negative binomial distributed. The objects of class *indivRisk* can be used as an input to the implemented local suppression functions. The local suppression function `localSupp2()` searches automatically for a quasi optimal solution to find a minimum amount of suppression. Additionally, the user may set weights for each variable to indicate which variables are less critical for suppression or which are more critical (the weight can be set to 0 for those variables which should not contain any suppressions and low weight for those which should only contain few suppressions). `localSupp2wrapper()` ensures to reach k-anonymity and low risk of re-identification.

It is recommended to use functions `freqCalc()`, `indivRisk()` and `globalRecode()` in an exploratory and sometimes in an iterative way. It takes just one look to know how many suppressions are necessary to achieve k-anonymity and/or low individual risk for different global recoding settings. In addition to that, one can simply reproduce any previous step of the anonymization. So, one can try out different global recodings of various variables and evaluate its impact on information loss. Moreover, the frequency count calculation procedure can deal with missing values in the data.

Function `pram()` provides the PRAM methodology (Kooiman et al. [2002]) and produces objects from class *pram*. When applying PRAM, values of categorical variables may change to different values, according to a pre-defined probability mechanism which is given in form of a specific transition matrix. A print method and a summary method are provided for objects of this class.

2.4.2 Methods for Masking Numerical Variables

The anonymization of numerical variables is often necessary to avoid a successful attack by record linkage methods.

Many different methods are implemented for masking numerical variables. For example there are more than 10 methods for microaggregation (*mdav* [Domingo-Ferrer and Mateo-Sanz, 2002], *rmd* (which is an extension to *mdav* proposed in Templ [2007a]), *individual ranking*, projection methods, and many more (see the references Templ [2006a], Anwar [1993], Elliot et al. [2005], Defays and Nanopoulos [1993], Defays and M.N. [1998], Domingo-Ferrer and Mateo-Sanz [2002]) in which the values of similar observations are aggregated by an "average" (which is often the arithmetic mean). In addition to that, five methods for adding noise are provided, namely, *adding additive noise* (the values of each variable are perturbed with a certain noise, see e.g. in Brand [2004]), *adding correlated noise* (the values of each variable are perturbed with a certain noise so that the covariance of these variables will be preserved, see Brand [2004]), *ROMM* (the perturbed data are obtained by $Y = AX$, whereas A is randomly generated and fulfills $A^{-1} = A^T$. To obtain an orthogonal matrix as described in Ting et al. [2005], the Gram-Schmidt procedure was chosen in package *sdcMicro*) as well as method *outdect* which evaluates the "outlyingness" of each observation and adds noise depending on the "outlyingness" of an observation (see e.g. in Templ [2006a]).

We also implemented numerical rank swapping (Dalenius and Reiss [1982]) which is a method based on sorting a variable by its values (ranking). Each ranked value is then swapped with another ranked value which had been randomly chosen within a restricted range, i.e. the rank of two swapped values cannot differ by more than p percent of the total number of values.

Methods `gadp()` and `shuffle()` are also implemented in *sdcMicro*). These methods were originally proposed by Muralidhar et al. [1999] where the model $Y = S\beta + \epsilon$ is the object of interest. Y are perturbed variables, S are non-confidential variables and $\epsilon \sim MVN(0, \Sigma_{XX} - \Sigma_{XS}(\Sigma_{SS})^{-1}\Sigma_{SX})$. The regression coefficients β are estimated by $\hat{\beta} = (S'S)^{-1}S'X$ using X, the confidential variables. Shuffling is then done by replacing the rank ordered values of Y (generated by GADP) with the rank ordered values of X. Please note, that these methods are not officially available on CRAN because these methods are under US-patent. Since these methods will only work reasonably when the data does not contain outliers, a robust version of these methods was proposed by Templ and Meindl [2008b] where robust Mahanlanobis distances are used to identify outliers and robust regression methods are applied. Since almost all data from official statistics contains outliers, it is highly recommended to use the robust version of shuffling.

There is also a method for fast generation of synthetic data included (details can be found in Mateo-Sanz et al. [2004a]) with which multivariate normal distributed data can be generated with respect to the covariance of the original data.

2.4.3 Data Sets Available in the Package

Several test data sets are included in the package. Some very small test data sets which were used by other authors for demonstration in the past are provided. Both the test data set from μ-ARGUS Hundepool et al. [2006] and the test data sets from the CASC project (Tarragona, Census and EIA data sets, see e.g. Hundepool [2004]) can be used.

2.4.4 New Methods

Several new algorithms for microaggregation are included and are also available in package *sdcMicro*. A simple approach is to cluster the data first and then to sort the data in each group by the most influential variables in the groups, or to sort the observations in each group by the first robust principal component obtained by projection pursuit (method `clustpppca` in package *sdcMicro*, see Templ [2006a] for details).

We have proposed a new algorithm for microaggregation called RMDM (**R**obust **M**ahalanobis **D**istance based **M**icroaggregation) where MDAV Domingo-Ferrer and Mateo-Sanz [2002] is adapted in the following way:

1. Compute the robust center of the data. This can be the L_1-median or the coordinate-wise median.

2. Consider the most distant observation x_r to the robust center using robust Mahalanobis distances. The MCD-Estimator (Rousseeuw P.J. [1999]) can be used to calculate the robust covariance matrix which is needed for the calculation of the robust Mahalanobis distances.

3. Find the most distant observation x_s of x_r using Euclidean distance.

4. Choose $k - 1$-nearest neighbors from x_r and also from x_s. Aggregate x_r and its $k - 1$ nearest neighbors with an average as well as x_s with their $k - 1$ nearest neighbors. The average can be the arithmetic mean but also a robust measure of location.

5. Take the previous dataset minus the aggregated data from the last step as a new dataset and continue with (1.) until all observations are microaggregated.

If the number of observations is not equal to a multiple of $2k$ then the algorithm aggregates the last remaining unaggregated $2k - h$, $h \in \{1, \ldots, k - 1\}$ observations with the chosen average (see also in Domingo-Ferrer and Mateo-Sanz [2002]). This proposed algorithm is more natural than the original MDAV algorithm since we deal with multivariate data taking the covariance structure of the data into account.

New methods for distance-based disclosure risk estimation for numerical data are proposed by Templ and Meindl [2008a] which consider again the "outlyingness" of observations. These methods are implemented as well as the methods proposed by Mateo-Sanz et al. [2004b]. The main difference to the methods of Mateo-Sanz et al. [2004b] is that depending on the "outlyingness" of an observation the disclosure risk interval around the original data or the masked data increases or decreases. This is a more natural approach since observations in the center of the data cloud generally bear a lower risk of re-identification than observations which are outliers.

A comparison of almost all methods can be found in Templ and Meindl [2008b] and can easily be derived by using package *sdcMicro*.

An overview of the methods implemented in the package is also given in Table 2.1 together with information about references and names of the methods (functions) in the package. There is also information about the computation time by using the μ-Argus data set with the same key variables as used in section 2.5. The rather subjective information of the complexity of the implementation may be informative for many researchers in this field.

Table 2.1: An overview of the most important methods which are implemented in sdcMicro. Note that for almost all of these methods *print*, *summary* and *plot* methods are implemented (and not listed here). Only those references to papers which have been used for the implementation of the methods in sdcMicro are given.

Method	Ref.	Name in sdcMicro	Param. of the function	Implem.	ctime
PRAM	Kooiman et al. [2002]	pram	–	easy	0.09
frequency calculation	Capobianchi et al. [2001]	freqCalc	–	depends	0.516
indivRisk	Franconi and Polettini [2004a]	indivRisk	–	medium	0.06
(optimal) local suppression	–	localSupp2-wrapper	–	hard	484
adding additive noise	Brand and Giessing [2002]	addNoise	additive	very easy	0.01
				Continued on next page	

Table 2.1 – continued from previous page

Method	Ref.	Name in sdcMicro	Param. of the function	Implem.	ctime
correlated noise	Brand and Giessing [2002]	addNoise	correlated	easy	0.01
correlated noise 2	Kim [1986], Hundepool et al. [2007b]	addNoise	correlated2	easy	0.01
restricted correlated noise	Brand [2004]	addNoise	restr	easy	0.01
ROMM	Ting et al. [2005]	addNoise	ROMM	easy	> 2000
adding noise based on multivariate outlier detection	Templ [2006a], here	addNoise	outdect	hard	0.18
rank swapping	Dalenius and Reiss [1982]	swappNum	–	very easy	0.88
microaggregation (ma), individual ranking	Defays and Nanopoulos [1993]	microagg.	onedims	easy	0.16
ma, sorting on first principal component	–	microagg.	pca	easy	0.1

Continued on next page

CHAPTER 2. SDC USING SDCMICRO

2.4. IMPLEMENTED METHODS AND DATA

Table 2.1 – continued from previous page

Method	Ref.	Name in sdcMicro	Param. of the function	Implem.	ctime
ma, sorting with projection pursuit pca on clustered data	Templ [2006a], Templ [2007a]	microagg.	clustpppca	hard	1.54
ma, mdav	Hundepool et al. [2007b]	microagg.	mdav	easy	> 200
ma, multivariate microaggregation based on robust Mahalanobis distances	Templ [2007a]	microagg.	rmd	hard	0.75
gadp	Muralidhar et al. [1999], Muralidhar and Sarathy [2006]	shuffle	–	very easy	0.16
shuffling	Muralidhar and Sarathy [2006]	shuffle	–	very easy	0.19
shuffling based on cgadp	Muralidhar and Sarathy [2006]	shuffle2	–	easy	0.19

Continued on next page

49

Table 2.1 – continued from previous page

Method	Ref.	Name in sdcMicro	Param. of the function	Implem.	ctime
robust gadp and robust shuffling	Templ and Meindl [2008b]	robGadp, robShuffle	–	medium	0.4
synthetic, cholesky decomposition	Mateo-Sanz et al. [2004a]	gendat	–	easy	0.01

2.5 A Small Tour in *sdcMicro*

To stay within the limit of pages only a small tour in *sdcMicro* can be done excluding most of the graphical results and some steps of recoding variables. Comments in the code are marked with #, the output from **R** with R>. For further details please have a look at the examples and documentation which are included in package *sdcMicro*.

Given the fact that you have already installed **R** one possibility of installing the *sdcMicro* package plus all required packages from a CRAN server is to type the following command in **R**

```
install.packages("sdcMicro", depend=TRUE)
```

Now you can load both the package and your data using the powerful **R** import and export facilities. For the sake of simplicity we use the μ-Argus test data and print the first eight columns (argument before] in the code below) of the first four observations (argument after []) of the data [4].

```
library(sdcMicro); data(free1); xtable(free1[1:4, 1:8])
```

	REGION	SEX	AGE	MARSTAT	KINDPERS	NUMYOUNG	NUMOLD
1	36	1	43	4	3	0	0
2	36	1	27	4	3	0	0
3	36	1	46	4	1	0	0
4	36	1	27	4	1	0	0

Please note that you can simply use your own more meaningful data. We just use this data set (partially) because many SDC-specialists are familiar with this data set and have tried out methods using this data within the μ-Argus software.

In the following code segment the frequency counts are calculated as described in Capobianchi et al. [2001] and allocated to object fr1 which is now automatically of class *freqCalc*. Several methods are available for this class. Object fr1 is then used as an input for the individual risk computation. A new object ir1 of class indivRisk will be produced. Several methods are available for this class as well. For example, function plot automatically knows which plot method must be used for an object of class *indivRisk*. Figure (2.4) is generated by this plot method. The implementation of this plot method for individual risk methods is quite similar to the one in μ-Argus.

Instead of providing metadata (which can be quite cumbersome) we only use the data itself. The user must only provide the information on which variables should be treated as the key variables and which one includes the sampling weights (if there is such a variable in the data). Since we neither have a detailed

---
[4]xtable() produces a LaTeX-styled print output.

variable description nor know much about the data (in practice this lack of information does not exists, because the data holder does have such information) we suppose that variables *REGION*, *SEX*, *AGE*, *MARSTAT*, *ETNI* and *EDUC1* may be treated as key variables because a data intruder might have some information about some observations of these variables. Please note that it is not possible to calculate these frequency counts in μ-Argus (it fails with an error message which reports that the program ran out of memory). Figure 2.3 shows an interactive plot of the individual risk similar to μ-Argus. Here, a histogram of the indidividual risk is shown but the empirical cummulative distribution function of the risk can be also plotted, which might be more informative.

```
kv <- which(colnames(free1) %in% c("REGION", "SEX",
                                    "MARSTAT", "ETNI", "EDUC1"))
fr1 <- freqCalc(free1, keyVars=kv, w=30)
rk1 <- indivRisk(fr1)
class(rk1)                    ## class of object rk1
R> [1] "indivRisk"
methods(class = indivRisk)    ## methods available for objects
                              ## of this class
R> [1] plot.indivRisk   print.indivRisk
plot(rk1)
```

The script shown above can easily be adapted (exclude the **R** results), e.g. by adding the function `globalRecode()`. `globalRecode()` recodes several categories of a variable into less categories or discretises a numerical variable. So, this function checks the class of a variable and recodes the variables based on its class. The user is still able to manipulate the data in an exploratory way. You might try to recode a variable in order to observe the influence of your recoding on the frequency count calculation and the individual risk computation, but also to find out how many suppression will be needed when changing the coding of some key variables. Note that you can easily reproduce all your steps either by

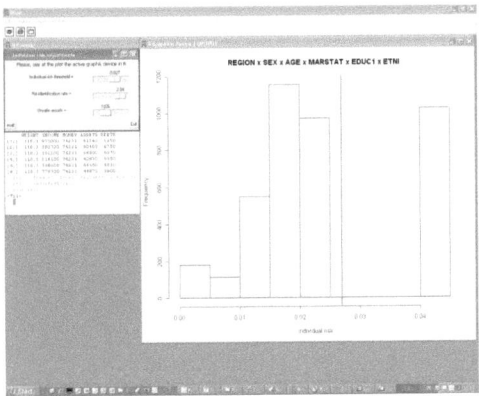

Figure 2.3: Individual risk. In the upper region of the figure you will see a helpful slider which is directly linked with the graphics.

running a part of your script or the whole script.

We now show just the code for the final recoding, namely recoding of region into 3 equally sized categories, recoding of *age* into 5 specific categories, recoding of education into 2 categories and recoding of ethnicity into 2 categories. This results in 54 observations which are unique in the sample (instead of 3702 without recoding). Since we do not have any information about these categories of this test data set, we only show a number of recodings which might not be meaningful. Naturally, the user has the information which category consists of which group, i.e. if education is equal to 9 means university degree or primary school, for example. With such information one may join the most similar categories into new categories. We are always able to check the effects of the recoding, e.g. showing how many observations are unique in the sample (fk=1).

```
x ← data.frame(free1)
attach(x)
```

```
x$REGION ← globalRecode(REGION, breaks=3, method="equalAmount")
x$AGE ← globalRecode(AGE, breaks=c(14,30,45,55,74))
table(EDUC1)
   1    2    3    4    5    6    7   9
 781  437  620  268 1223  457  209   5
x$EDUC1 ← globalRecode(EDUC1, breaks=c(0,4,9))
table(ENTI)
    1    2    3    4    5    6    8    9
 3677   23   19   13   73   79   55   10   51
x$ETNI ← globalRecode(ETNI, breaks=c(0,1,9))
fr2 ← freqCalc(x, keyVars=c(1:4,9,11), w=30)
fr2


54 observation with fk=1
63 observation with fk=2


rk2 ← indivRisk(fr1)
detach(x)
```

It is really easy to see the effects of different recodings in less than 0.5 seconds, for example, when allowing more regions (e.g. 5). Just type the information about the new breaks into this script and run the code resulting in

```
fr2


107 observation with fk=1
110 observation with fk=2
```

After minimising the re-identification risk you can apply local suppression (function `localSupp2()`) on object `indivf` and `fr1` to delete the last unsureness about risky observations and then make use of the implemented print and summary

CHAPTER 2. SDC USING SDCMICRO

2.5. A SMALL TOUR IN SDCMICRO

methods. But global recoding and local suppression may, of course, also be used in an alternated manner. With parameter *importance* in function *localSupp2Wrapper* weights to each key variable can be assigned. Variables with higher weights will be preferred for suppression. In our example we assume that suppressions in the last 3 key variables are more acceptable for the user than in the other key variables. We want to perform 3-anonymity.

```
ls1 ← localSupp2Wrapper(x,keyVars=kv,w=30,importance=c(0.2,0.2,0.3,1,1,1),kAnon=
ls1

[1] "Total Suppressions in the key variables 280"
[1] "Number of suppressions in the key variables "

4 0 0 202 54 20

[1] "3-anonymity == TRUE"
```

So, 4 values in *REGION* are supressed and 202, 54, and 20 values are suppressed in *MARSTAT*, *EDUC1* and in variable *ETNI* respectively.

Finally, one can check if observations with high individual risk remain and may suppress some additional values (for observations with a risk above a certain threshold) in order to reduce the risk (using function *localSupp*). This can easily be done by

```
fff ← freqCalc(ls1$xAnon, keyVars=c(1:4,9,11), w=30)
rk11 ← indivRisk(fff)
plot(rk11)
```

In addition to that, you can simply microaggregate numerical variables with more than 10 different methods or use another method described previously.

```
x[, 31:34] ← microaggregation(x[, 31:34], aggr=4, method="clustpppca")
```

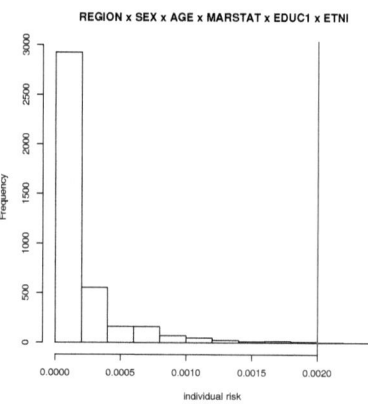

Figure 2.4: Individual risk after the anonymization of the μ-Argus test data set.
There are a lot of comparison plot methods available to compare the perturbed data and the original data. You can easily compare the different methods. We show this on another data set, the *Tarragona* data set[5], because the μ-Argus test data does include faked numerical variables which follow a multivariate uniform structure without correlations between the variables.

Evaluating different methods for the masking of numerical variables becomes easy by using function `valTable()` which estimates different measures of information loss and distance-based disclosure risk.

```
data(Tarragona)
v ← valTable(Tarragona, method=c("addNoise: additive", "addNoise: correlated2",
"addNoise: outdect","swappNum","onedims","pca","clustpppca","mdav","rmd",
"dataGen"))
xtable(v[, c(1,3,6,9,15,17,20)])   ## choose some measures and produce Table 2
```

[5]http://neon.vb.cbs.nl/casc/testsets.html

Table 2.2: Comparison of different methods regarding a univariate measure of information loss (mean absolute error of medians), one multivariate measure of information loss (mean absolute error of correlations), one utility measure, the risk measure given in Mateo-Sanz et al. [2004b] and a new risk measure weighted with the robust Mahalanobis distance. Further measures were evaluated but not printed in this example.

	method	amedian	amad	acors	util1	risk0	risk2
1	addNoise: additive	0.23	0.16	0.18	9.61	0.04	0.99
2	addNoise: correlated2	3.57	5.88	1.20	93.96	0.00	1.00
3	addNoise: outdect	0.26	0.24	0.23	14.99	0.59	0.99
4	swappNum	0.13	0.18	1.28	386.62	0.00	0.99
5	onedims	0.03	0.04	0.01	18.63	0.80	0.99
6	pca	2.62	1.10	9.57	232.68	0.00	0.00
7	clustpppca	3.64	1.55	7.40	259.34	0.00	0.00
8	mdav	1.98	0.83	5.68	197.10	0.00	0.00
9	rmd	0.83	0.42	1.48	170.96	0.00	0.00
10	dataGen	27.53	78.48	5.89	1162.14	0.00	0.83

The first column of Table 2.2 represents a univariate measure of information loss (absolute deviations from the medians), the next column represents a multivariate measure of information loss (differences in spearman correlation coefficients) followed by a data utility measure and two distance-based disclosure risk measures. A detailed description of these measures can be found in the online-help files from package *sdcMicro* (Templ [2007d]). One can easily compare the performace of these methods on the Tarragona data set. Method adding additive noise performs quite well but bears a high disclosure risk, for example. However, Table 2.2 shows only one configuration and naturally for each method some parameters can be changed so that the data utility increases and the disclosure risk decreases or vice versa. Methods `shuffle()` and `gadp()` are not included in Table 2.2 since it would be rather subjective which variables should be chosen to be the "non-confidential" ones to protect these variables. In many cases method RMDM (in package *sdcMicro* the method is named `rmd`) performs best since it results in relatively low disclosure risk and very low information loss.

Another method for categorical variables is PRAM which can easily be applied with function `pram()`. In the following, we perturbe variable MARSTAT form the μ-Argus test data set with the invariant PRAM methodology. A lot of information is stored in object MARSTATpram, e.g. the invariant transition matrix. Summary and print methods are provided as well. A selection of the output of the summary is given below.

```
MARSTATpram ← pram(free1[,"MARSTAT"])    ## with default parameters
summary(MARSTATpram)


original frequencies              transition Frequency
 1     2     3     4            1    1 ---> 1      2448
2547  162   171  1120           2    1 ---> 2        27
                                3    1 ---> 3        28
                                4    1 ---> 4        44
frequencies after perturb.:     5    2 ---> 1        33
                                6    2 ---> 2       118
 1     2     3     4            7    2 ---> 3         4
2571  160   178  1091           8    2 ---> 4         7
                                9    3 ---> 1        20
                               10    3 ---> 2         3
                               11    3 ---> 3       130
```

2.6 Open Source Initiative

As mentioned above, the entire code is freely available and can be downloaded from
http://cran.r-project.org. So you can learn from this code, change it yourself or develop the package *sdcMicro* further. Instead of keeping the developed code to yourself you are invited to contribute to this package. Every response and bug report will be helpful in achieving a higher quality of the package. Note that up to now the quality of the package was highly improved by comments and bug reports from many users from statistical offices and companies. *sdcMicro* is actively being developed and improved, and so it is highly recommended to use the latest version of the package.

Every function has its own author and the copyright of the function belongs to the author. This means that nobody can use your functions for a commercial software product. On the other hand the intellectual property is also ensured. Please note that the intellectual property must be guaranteed when using the functions for publications in journals (please reference when using *sdcMicro* or parts of the package). But at the same time all the functions are open-source and everybody can, and should, use them.

2.7 Conclusion

The main difference between *sdcMicro* and the popular software μ-Argus is the possibility to use SDC methods in an exploratory manner in *sdcMicro*. So, different parameter settings can be tried out during the data masking process and a detailed look at each step of the anonymization using implemented print, summary, plot, information loss and disclosure risk methods is at hand. Additionally, the whole functionality of **R** can be used. All the steps of the anonymization can easily be reproduced.

We have shown by anonymizing the μ-Argus test data set that the anonymization procedure becomes easy using **R**-package *sdcMicro*. Moreover, for your own data sets it might get even simpler because detailed information about the variables and its categories is at hand.

The potential of this package can become very high and the package has a realistic chance to become the most important implementation for SDC in microdata protection not only because all methods from μ-Argus are available but also because it features several new developments. The response of users from all over the world is very positive and the package is already used in the production process (see e.g. Meindl and Templ [2007]). The users are still satisfied with the command line interface, and so an implementation of a graphical user interface has not been developed yet. In addition to this package, the entire power of **R** can be used to boost the results. Everybody is invited to contribute to this package, especially in funded future research projects.

Acknowledgements

My thanks go to Bernhard Meindl for numerous suggestions and contributions to the code and to Alois Haslinger who supported the development of the package by giving me the necessary time for this work.

3 Robustification of Microdata Masking Methods and the Comparison with Existing Methods

Published in "Privacy in Statistical Databases. Lecture Notes in Computer Science", Springer [Templ and Meindl, 2008b]

Matthias Templ[*,**], Bernhard Meindl[*],

[*] Department of Methodology, Statistics Austria, Guglgasse 13, 1110 Vienna, Austria. (bernhard.meindl@statistik.gv.at) and

[**] Department of Statistics and Probability Theory, Vienna University of Technology, Wiedner Hauptstr. 8-10, 1040 Vienna, Austria. (templ@statistik.tuwien.ac.at)

Abstract:. The aim of this study was to compare different microdata protection methods for numerical variables under various conditions. Most of the methods used in this paper have been implemented in the R-package *sdcMicro* which is available for free on the comprehensive R archive network (http://cran.r-project.org). The other methods used can be easily applied using other R-packages. While most methods work well for homogeneous data sets, some methods fail completely when confidential variables contain outliers which is almost always the case with data from official statistics. To overcome these problems we have robustified popular methods such as microaggregation or shuffling which is based on a regression model. All methods have beed tested on bivariate data sets featuring different outlier scenarios. Additionally, a simulation study was performed.

Keywords: Statistical disclosure control, numerical data protection methods, robustness, simulation.

3.1 Introduction

One of the most important steps in the process of data anonymization is to anonymize the categorical variables. This means to anonymize indirect identifiers with respect to the sampling weights.

However, an attacker may try to identify statistical units using numerical variables by using linking and/or matching procedures as well. The anonymisation of numerical variables makes it more difficult for the attacker to successfully match or merge underlying data with other data sources. Therefore, the anonymisation of numerical variables is of high interest too.

Please note that R-package *sdcMicro* contains a large amount of methods for the protection of categorical data (see, e.g., Templ [2007b] and Meindl and Templ [2007] for a practical application). In the following work we will focus on methods for anonymizing numerical variables such as microaggregation, adding noise, swapping and shuffling.

Serveral perturbation methods for numerical data (microaggregation via z-transformation, rank swapping, resampling, generation of synthetic data based on samples that are generated from the empirical mean and covariance structure of the data) have been compared within a simulation framework using bivariate data, for example, by Karr et al. [2006].

However, Templ [2006a] notes that better microaggregation procedures exist. Furthermore, rank swapping destroys the multivariate data structure (Templ [2006a]). Synthetic data are approximately normal distributed but the original data do not follow and can not be transformed to follow a normal distribution in general.

Another simulation study was carried out in Muralidhar et al. [2006] where swapping was compared to shuffling on data sets generated from bivariate normal distributions featuring different correlations. A similar simulation study was conducted in Muralidhar and Sarathy [2006] where the distribution of the confidential variables followed other theoretical distributions, but no outliers were included.

The results of Karr et al. [2006] are based on random data sampled from theoretical distributions without contamination. But in real world applications we often can not assume that the data follow a theoretical distribution. Therefore, perturbation methods must also give reasonable results when outlier exist in the data.

In the next section we introduce the methods that we have evaluated. Details can be found in the references and the implementation can be found in the R-package *sdcMicro* (Templ [2008d], Templ [2007a], Templ [2007b]) which can be downloaded on the comprehensive R archive network (CRAN, http://cran.r-project.org, see R Development Core Team [2008a]).

3.1.1 Methods Used in this Study

We now give an overview on the methods that have been investigated. The methods can be classified into four groups: methods based on sorting, methods based on grouping, methods based on adding noise and methods for synthetic data generation.

In Table 3.1 we give references for each method as well as the the name of the corresponding function in R-package *sdcMicro*. Some functions (e.g. addNoise, microaggregation) are wrapper functions for several methods which are listed in Table 3.1 as well. Furthermore, the parameters used in this study are given in the

CHAPTER 3. ROBUSTIFICATION OF MASKING METHODS

3.1. INTRODUCTION

last column of the table since they are different from the default ones. The parameters and their default choices are described in detail in the package manual of *sdcMicro*. Please note that more method especially on synthetic data generation are implemented in *sdcMicro*.

Table 3.1: Investigated methods. Only those references to papers are listed which have been used for the (re-)implementation of the methods in *sdcMicro*.

Method descrip.	Reference	Function in sdcMicro	Method	Param. used
adding additive noise	Brand and Giessing [2002]	addNoise	additive	noise=200
correlated noise	Brand and Giessing [2002]	addNoise	correlated	noise=200
correlated noise 2	Kim [1986], Hundepool et al. [2007b]	addNoise	correlated2	noise=200
restricted correlated noise	Brand [2004]	addNoise	restr	noise=200
ROMM	Ting et al. [2005]	addNoise	ROMM	noise= $200, p = 0.01$
adding noise based on multivariate outlier detection	Templ [2006a]	addNoise	outdect	noise=200
				Continued on next page

CHAPTER 3. ROBUSTIFICATION OF MASKING METHODS
3.1. INTRODUCTION

Table 3.1 – continued from previous page

Method descrip.	Reference	Function in *sdcMicro*	Method	Param. used
rank swapping	Dalenius and Reiss [1982]	swappNum	-	p=15, p=40
ma, sort on a single variable	-	microaggregation	single	default
ma, individual ranking	Defays and Nanopoulos [1993]	microaggregation	onedims	default
ma, influence	Templ [2006a]	microaggregation	influence	default
ma, sorting on first principal component	-	microaggregation	pca	default
ma, sorting with projection pursuit pca	Templ [2006a], Templ [2007a]	microaggregation	pppca	default
ma, sorting with projection pursuit pca on clustered data	Templ [2006a], Templ [2007a]	microaggregation	clustpppca	default
ma, mdav	Hundepool et al. [2007b]	microaggregation	mdav	default
			Continued on next page	

CHAPTER 3. ROBUSTIFICATION OF MASKING METHODS
3.1. INTRODUCTION

Table 3.1 – continued from previous page

Method descrip.	Reference	Function in *sd-cMicro*	Method	Param. used
ma, multivariate microaggregation based on robust Mahalanobis distances	Templ [2007a]	microaggregation	rmd	default
gadp	Muralidhar et al. [1999], Muralidhar and Sarathy [2006]	shuffle (output *gadp*)	gadp	default
shuffling	Muralidhar and Sarathy [2006]	shuffle (output *shuffle*)	shuffle	default
shuffling based on cgadp	Muralidhar and Sarathy [2006]	shuffle2	shuffle2	default
robust gadp	this paper	robShuffle (output *gadp*)	robGadp	default
robust shuffling	this paper	robShuffle (output *shuffle*)	robShuffle	default

Methods based on adding noise: One possible procedure is to add additive

66

noise (cited as *noise* in Tables and Figures) to each numerical variable

$$Y = X + \epsilon ,$$

where $X \sim (\mu, \Sigma), \epsilon \sim N(0, \Sigma_\epsilon)$, $\Sigma_\epsilon = \alpha \cdot diag(\sigma_1^2, \sigma_2^2, \ldots, \sigma_p^2), \alpha > 0$, $Cov(\epsilon_i \neq \epsilon_j) \; \forall i \neq j$ and p is equal the dimension of the numerical variables which should be perturbed (see e.g. also in Brand and Giessing [2002]).

It is clear that multivariate measures such as the correlation coefficient can not be preserved after adding additive (uncorrelated) noise. Correlation coefficients can however be preserved when correlated noise is added. In this case the covariance matrix of the masked data is $\Sigma_Y = (1+\alpha)\Sigma_X$ (see e.g. in Brand and Giessing [2002]).

For method *correlated2* $d = \epsilon(1 - \alpha^2)$ and then $x_j d + \alpha z_j$ is calculated where z_j are random numbers drawn from $N(\frac{(1-d)\bar{x}_j}{\alpha}, s_j)$ with s_j being the standard deviation of X_j (see e.g. in Kim [1986] or Hundepool et al. [2007b]).

The restricted correlated noise method (implemented as method *restr* in sdcMicro) is a similar method that takes the sample size into account (Brand [2004]).

Furthermore, method ROMM (Random Orthogonal Matrix Masking, Ting et al. [2005]) has been considered in the simulation study. In this method perturbed data are obtained by the transformation $Y = AX$ where A is randomly generated and fulfills the orthogonality condition $A^{-1} = A^T$. To obtain a orthogonal matrix as described in Ting et al. [2005] the Gram-Schmidt procedure was chosen in the computational implementation of method *ROMM*.

In order to be able to deal with inhomogeneous data sets including outliers the first author of the paper has implemented a method in which outliers are detected. Observations with large robust Mahalanobis distances are treated as outliers as well as observations that exhibit univariate outliers, i.e. where the values of a variable x are greater than robust measure of location (e.g. the median) plus a robust measure of scatter (usually the Median Absolute Deviation (see Huber

CHAPTER 3. ROBUSTIFICATION OF MASKING METHODS
3.1. INTRODUCTION

[1981])).

Outliers should be protected more (by default by adding additive noise) than the rest of the observations because outliers show a higher risk for re-identification than non-outliers. This method is denoted *outdect* in package *sdcMicro* and in the following text.

Methods based on sorting: Rank swapping (see Dalenius and Reiss [1982] and Moore [1996]) is a method based on sorting a variable by their numerical values (ranking). Each ranked value is then swapped with another ranked value that has been chosen randomly within a restricted range. This means for example that the rank of two swapped values cannot differ by more than p percent of the total number of observations. Rank swapping must be applied to each variable separately.

Methods based on sorting and grouping: A familiar definition of microaggregation can be found at http://neon.vb.cbs.nl/casc/Glossary.htm: "Records are grouped based on a proximity measure of variables of interest, and the same small groups of records are used in calculating aggregates for those variables. The aggregates are released instead of the individual record values."

The choice of the "proximity" measure is the most challenging and most important part in microaggregation since the multivariate structure of the data is only preserved if similar observations are aggregated. Sorting data based on one single variable in ascending or descending order (method *single* in *sdcMicro*), sorting the observations in each cluster (after clustering the data) by the most influencial variable in each cluster (method *influence*, see Templ [2006a]) as well as sorting (and re-ordering the data after aggregation) in each variable (individual ranking method, see Defays and Nanopoulos [1993]) is not optimal for multivariate data (see Templ [2007a]).

Projection methods which sort the data according to the first principal component (method *pca*) or its robust counterpart (method *pppca*, see Templ [2006a])

CHAPTER 3. ROBUSTIFICATION OF MASKING METHODS
3.1. INTRODUCTION

can be improved when these methods are applied to clustered data (method *clustpppca*, see Templ [2007b], for example). In order to estimate the first principal component in a robust way, it is necessary to obtain a a robust estimate of the covariance matrix. However, this is only feasible for small or medium sized data sets using methods like M-estimation Maronna [1976], the MVE estimator Rousseeuw [1985] or the orthogonalized Gnanadesikan-Kettering (OGK) estimator Maronna and Zamar [2002]. Furthermore, all principal components must be estimated when using classical approaches for PCA. Method *pppca* avoids this and estimates the first (robust) principal component without the need of estimating the covariance.

The Maximum Distance to Average Vector (MDAV) method is an evolution of the multivariate fixed-size microaggregation (see Domingo-Ferrer and Mateo-Sanz [2002], for example). This method (*mdav* in *sdcMicro*) is based on Euclidean distances in a multivariate space.

The algorithm has been improved by the first author of the paper by replacing Euclidean distances with robust Mahalanobis distances. In Templ [2007b] a new algorithm called RMDM (**R**obust **M**ahalanobis **D**istance based **M**icroaggregation) was proposed for microaggregation where MDAV Domingo-Ferrer and Mateo-Sanz [2002] is adapted in several ways. The proposed procedure (details can be found in Templ [2007a]) is a more natural approach than MDAV since multivariate data are dealt with by taking the (robust) covariance structure of the data into account.

Methods based on models: GADP (method *gadp* in *sdcMicro*) is based on the model $Y = S\beta + \epsilon$ and was originally proposed by Muralidhar et al. [1999]. Y are perturbed variables, S are non-confidential variables and $\epsilon \sim MVN(0, \Sigma_{XX} - \Sigma_{XS}(\Sigma_{SS})^{-1}\Sigma_{SX})$. The regression coefficients β are estimated by $\hat{\beta} = (S'S)^{-1}S'X$ using X, the confidential variables. The procedure is described in detail in http://gatton.uky.edu/faculty/muralidhar/CDAC42.ppt.

CHAPTER 3. ROBUSTIFICATION OF MASKING METHODS

3.1. INTRODUCTION

Since correlations between S and X have to be estimated in the procedure, Muralidhar and Sarathy [2006] refers to choose the Spearman rank correlation measure when dealing with non-normal distributed data. Shuffling is finally done by replacing the rank ordered values of Y (generated by GADP) with the rank ordered values of X.

Another approach is the copula based GADP which is described in Muralidhar and Sarathy [2006] and which was investigated too. It is assumed that the data follow a theoretical distribution from which the inverse distribution function can be expressed in analytical form. This approach is only feasible if the data follow a theoretical distribution approximately. However, in real world data sets this assumption can hardly be justified since data almost always include outliers. Therefore, similar conclusions as in the GADP approach have been obtained since this copula based approach cannot deal with outlying observations.

These very simple methods are under US-Patent (7200757) and so the re-implementation of the method is not included in package *sdcMicro*. Nevertheless, the scientific work must go on and therefore the first author of the paper has implemented a - not patented - extension of the procedure which can deal with outliers as well (see section 3.2).

Synthetic data: In package *sdcMicro* a method based on the Cholesky decomposition for fast generation of synthetic data has been re-implemented with which multivariate normal distributed data can be generated with respect to the covariance structure of the original data (details can be found in Mateo-Sanz et al. [2004a]). However, this does not reflect the distribution of real complex data since such data are generally not multivariate normal distributed and include outliers. Thus, this method was not considered in the simulation study.

Another method which is based on regression is called the IPSO synthetic data generators (see Burridge [2003] or Torra et al. [2006] for an application). However, the generation of synthetic data using regression models will fail when dealing with inhomogeneous data sets including outliers. This method gives reasonable results only for data that are multivariate normal distributed. For this reason, this method is not further addressed in this paper.

A well known method is blanking and imputation (Griffin et al. [1989]). Since outlying observations may be identified easier than non-outliers it is clear that such a method must blank and impute the outliers in any case. However, an imputation of outliers without destroying the multivariate structure of the data is difficult.

Rubin [1993] suggested to generate a completely synthetic data set based on the original survey data and multiple imputation. This method is not addressed in this simulation study as well.

Latin Hypercube Sampling (Iman and Conover [1982], Stein [1987], Wyss and Jorgensen [1998]) was also not considered in the paper because this method gives worst results and we are not sure if this results from a mistake in our re-implementation of the method. Unfortunately, also the results based on a MAT-LAB code from another author are worst (Minasny [2003]). This is not a surprise because the inverse of the distribution function must be available for this method. An analytical form of the inverse of the distribution function is rather complex or impossible to find for inhomogeneous data sets.

3.1.2 Information Loss

First we want to overview existing measures of information loss which are almost always univariate measures. In this paper we concentrate on multivariate measures of information loss which evaluate the multivariate structure of the original and the perturbed data.

One measure of information loss, called IL1s (see e.g. in Yancey et al. [2002] or Mateo-Sanz et al. [2004b]) and is based on aggregated distances from original data points to corresponding values from the perturbed data divided by the

CHAPTER 3. ROBUSTIFICATION OF MASKING METHODS
3.1. INTRODUCTION

standard deviation for each variable. Unfortunately, this measure is large even if only one outlier was perturbed highly and all values are exactly the same as in the original data set.

Other measures are considered in Hundepool et al. [2007b] and Templ [2006a]. Measures of information loss which compare univariate statistics of the original data and the perturbed data are for example the sum of the differences of the mean or medians. Measures which compare multivariate statistics of the original data and the perturbed data evaluate differences of the correlation matrices or loadings in principal component analysis. In our study we compare the eigenvalues of the classical covariance and the robust covariance Rousseeuw [1985] of the original data with the ones from the perturbed data. Other kinds of measures of information loss are discussed in Karr et al. [2006].

Improved measures of information loss has been suggested by Mateo-Sanz et al. [2005] and are also implemented in *sdcMicro* as well as robust measures (see in the package manual of *sdcMicro*).

3.1.3 Disclosure Risk

In Domingo-Ferrer et al. [2001] a measure of disclosure risk is proposed which is based on distances and assumes that an intruder has additional information (disclosure scenarios) so that he can link the masked record of an observation to its original value (see e.g. Mateo-Sanz et al. [2004b]). Given the value of a masked variable it is checked whether the corresponding original value falls within an interval centered on the masked value. The width of the interval is based on the rank of the variable or on its standard deviation Domingo-Ferrer et al. [2001]. However, this interval does not depend on the scale of the actual value and therefore the length of the interval is equal for non-outlying and outlying values. However, outlying observations should be much more perturbed than non-outliers.

Another type of measures of disclosure risk - the value disclosure risk - is extensively used, e.g. by Muralidhar et al. [2006]. The main goal of this measure is to evaluate the gain in explanation of parameters or variables when perturbed data are released.

Templ and Meindl [2008a] suggests a new and more realistic measures of disclosure risk which accounts for outlying observations by using robust Mahalanobis distances. The robustification was done using the MCD-estimator Rousseeuw [1985].

3.2 Robustification of GADP and Shuffling

When classical methods are fitted to data which include outliers one may get unrealistic estimation results. On the other hand robust methods are able to describe the majority of the data and to detect outliers. In the context of gadp and shuffling it is then possible to perturbate this majority of the data reasonable but it is not possible to perturbe or generate the outliers in a reasonable way. Therefore high data utiltiy cannot be reached with non-robust nor with robust procedures. Therefore, we propose to use robust procedure both for the non-outliers and for the outliers seperately. Please note that when using mixture models to describe the data the same problems with outliers occur.

It is easy to see that classical shuffling procedures are influenced by outlying observations because within the procedure a rank based covariance matrix estimation and a least squares regression fitting is applied.

In order to robustify shuffling one has to modify the following conditions:

- Choose a robust regression method with a high breakdown point instead of least squares estimation. The *breakdown point* of an estimator measures the maximal percentage of the data points which may have been contaminated before the estimates become completely corrupted.

- Choose a robust estimator with a high breakdown point to estimate the correlation between the confidential and the non-confidential variables.
- Define outliers to be the observations with robust Malahanobis distances larger than $\sqrt{\chi^2_{(0.975,\ p)}}$ with p being the number of variables.
- Apply shuffling to non-outlying observations.
- Apply another perturbation method to the outlying observations.

In Figure 3.1 as well as in Figure 3.2 it is shown that shuffling (and therefore gadp as well) will be seriously influenced by o utliers while the robust shuffling procedure gives reasonable results (Fig. 3.2; we had applied RMDM microaggregation for the outlier part) also when the data are contaminated.

3.3 Results based on specific artificial data sets

Most of the methods under consideration were already evaluated and compared based on real data in Templ [2006a], Templ [2007b] and Templ [2007a] (see also in the R online help files of package *sdcMicro* Templ [2008d]). In this section we will investigate artificial data sets featuring different outlier scenarios.

Each method was applied to several bivariate data sets which are visualized at the top of Table 3.2. The first two data sets follow a bivariate normal distribution with uncorrelated (first data set) and correlated variables (second data set). The only difference between data sets 1 and 2 and data sets 3 and 4 is the inclusion of a single outlier. Furthermore we test all methods on a data set that features an outlier group.

Table 3.2 provides detailed information about the performance of the methods under consideration with respect to data utility and data protection. Columns *prot.* in Table 3.2 indicate the performance of the microdata protection methods regarding data protection while columns *qual.* show the performance of the methods

CHAPTER 3. ROBUSTIFICATION OF MASKING METHODS
3.3. RESULTS BASED ON SPECIFIC ARTIFICIAL DATA SETS

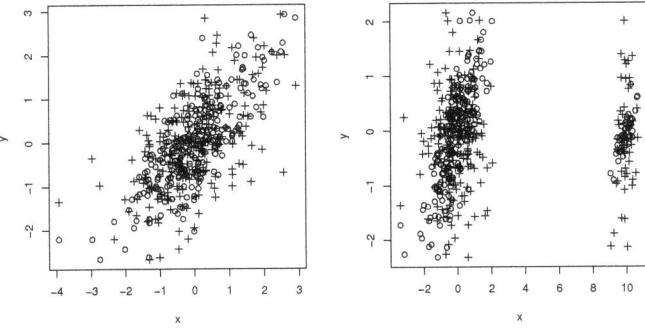

Figure 3.1: LEFT: The original bivariate normal distributed data (circles) and the shuffled data (crosses) do have a quite similar behavior. RIGHT: The original data consists of bivariate normal distributed data (non-outliers) plus a shifted outlier group and the shuffled data (crosses) show a quite dissimilar behavior.

with respect to the fact whether the original data structure had been destroyed after applying a procedure.

Possible values of these columns are *pass*, *part.* and *fail* which mean that the method passes, partly passes or fails the criteria. Regarding data utility we simple look at the bivariate original data and compare it to the masked data similar to the evaluation of shuffling in 3.1 and 3.2. When evaluating disclosure risk we pay special attention if and how the data are perturbed, especially if the outliers are protected sufficiently.

The classification of the masking procedure given a certain data scenario into *pass*, *part.* and *fail* is rather subjective but more powerful than using one measure of information loss or a traditional measure of disclosure risk.

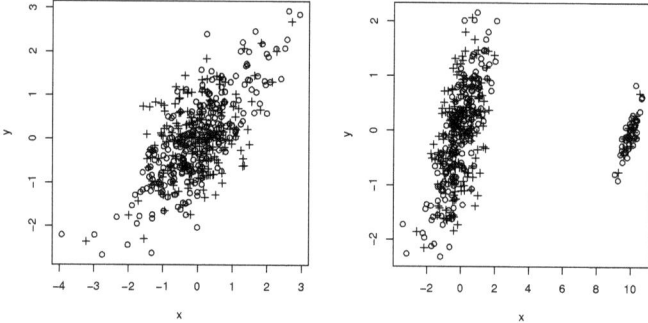

Figure 3.2: LEFT: The original bivariate normal distributed data (circles) and the shuffled data (crosses) do have a quite similar behavior. RIGHT: The original data consists of bivariate normal distributed data (non-outliers) plus a shifted outlier group and the shuffled data (crosses) also show a quite similar behavior.

This fact becomes clear when thinking of the evaluation of disclosure risk regarding to the third graphic in Figure 3.2. If a method only fails to protect the outlier a usual global measure of disclosure risk (for example measure SDID, see Mateo-Sanz et al. [2004b]) would be very low. However, the protection of the probably most interesting observation for a data intruder is essential.

Furthermore, when using certain measures the evaluation of the methods depends on the chosen measures and therefore we prefer the explorative approach of visually assessing the performance of the masking methods.

Table 3.2 shows that almost all methods work well, providing good quality and data utility when the procedures are applied to normally distributed data featuring different covariances. If outliers are included in the data most of the classical methods have problems with respect to data utility while using robust procedures avoid these problems. Using shuffling it is not possible to protect the single outlier in graphic 3 and 4 since the outlier of the generated data (e.g. with method gadp) has the same rank as the outlier of the original data for sure and the swapped value is exactly the same as the original value. Thus it is not possible to perturbe such an outlier using method *shuffle* while the robust version avoids this problem.

When working with real complex data a minimum requirement of a method is that they should only feature *pass* in Table 3.2. Hence, only methods *clustppca*, *mdav*, *rmd*, *outdect*, *robShuffle* and *robGapdp* should be applied on real data sets.

3.4 Simulation

3.4.1 Design of the Simulation Study

As already indicated, all methods discussed above have been applied to synthetic, bivariate data sets that have been sampled from a multivariate normal distribution with mean vector $\mu = (0,0)$ and covariance matrix Σ ($\sigma_{ii} = (1,1), \sigma_{ij} = (0.8, 0.8)$). The data sets that have been used to assess the quality of the microaggregation procedures feature different proportions of outliers (from 0% up to 40%). In the simulation study we considered shifted outliers that have also been generated by a multivariate normal distribution. However, the mean vector of the outlying observations is different ($\mu_{out} = (10, 0)$) to the mean vector of the non-outliers. Furthermore, different correlations between the variables have been considered by adjusting the covariance matrix.

We also assess the stability of the different microaggregation methods with respect to measures for data utility and risk. Thus, a total number of 300 data sets was generated for each combination of outlier percentage and correlation between the two variables. All methods were then applied to all data sets featuring a given correlation and outlier percentage. Analyzing the results it is possible to discuss which methods are providing stable outcomes in terms of data quality (protection) and data utility.

3.4.2 Simulation Results

The following results are based on a simulation study using a total of 1000 simulation runs. The results show again how the existence of outliers influences some microprotection procedures. All the following graphs show the median of the 1000 simulation runs given the outlier fraction.

The first results (displayed in the graphics at the top of Figure 3.3) are summarizing results of information loss and disclosure risk measures which are based on distances between the perturbed and the original data. The graphics at the bottom of Figure 3.3 summarize the simulation results based of an information loss criteria which is based on absolute differences between the eigenvalues of the robust covariance matrices in the original and the perturbed data and disclosure risk criteria which is based on robust Mahalanobis distances and neighbourhood comparisons (see Templ and Meindl [2008a]).

The graphic at the top left of Figure 3.3 shows the influence of outliers to various methods based on the IL1 information loss measure. This measure does not evaluate how well certain statistics are preserved. It only evaluates distances between the original and the preserved data. Therefore, *shuffling* and *gadp* exhibit the highest "information loss" and low "disclosure risk". If the amount of outliers in the data increases, also the information loss criteria shows higher values. This indicates the influence of outliers to these methods. This is not the case for *robust shuffling*. It is also clearly visible that the *RMDM* (denoted as *rmd*) method performs better than *mdav*. Naturally, these two methods have high "risk of disclosure" since only distances are evaluated but the advantages of microaggregation

- the aggregation of observations which provides a good protection by itself - is not considered in this risk measure.

The simulation results shown in the bottom of Figure 3.3 are based on more realistic measures of information loss and disclosure risk. These measures take the multivariate behavior and the risk with respect to robust Mahalanobis distances into account. Microaggregation methods *rmd* and *mdav* perform again very well and also *robust shuffling* gives quite good results. *Shuffling*, *ROMM*, *gadp* and *adding noise* are highly influenced by outliers and perform poorly. The "shift" between 0 percent outliers and 2.5% outliers in the graphic at the bottom right of Figure 3.3 occurs for many methods. This "shift" indicates that even a few outliers have influence on these methods which results in poor perturbation quality if the data contains outliers. It is self-evident that poor results regarding information loss often indicate low disclosure risk, i.e. if the (multivariate) structure of the data is completely destroyed not much can be uncovered by an intruder.

3.5 Conclusion

We have conducted a large simulation study considering various outlier scenarios and different correlations between the variables. Only a few results in a very comprehensive form could be presented in this paper in order to stay within the limit of pages.

Outliers are virtually present in every data set from official statistics and therefore perturbation methods for numerical variables must be able to deal with inhomogeneous data sets. Furthermore, outlying observations surely possess a higher risk for re-identification and it is essential that the methods protect these outliers properly. On the other hand, a protection method should not destroy the multivariate data structure. We showed that many methods are heavily influenced by outliers which results in poor quality regarding data utility and protection.

We showed that some classical methods are not able to deal with special data configurations (see Table 3.2). These methods may not be suitable for applications to real world data sets. Nevertheless, some of the most popular methods which fail under such data configurations have been included in the simulation study together with robust modifications of these methods. However, some methods (mostly methods for synthetic data generation) were excluded from the simulation study because it is clear that these methods can not deal with data that feature outliers.

The results of the simulation study showed some procedures performed poorly when applied to data that are contaminated with outliers. Analyzing the results it turned out that methods *rmd*, *clustpppca*, *mdav* and *robust shuffling* performed very well. In many situations *rmd* outperformed all other methods (see Fig. 3.3).

With model based procedures one may run into serious problems when masking real complex data including outliers. The robustification of such methods, like our proposed robust shuffling procedure, makes it possible to deal with data contamination in an efficient way.

CHAPTER 3. ROBUSTIFICATION OF MASKING METHODS

3.5. CONCLUSION

Table 3.2: Detailed check if the methods can deal with special data configurations, i.e. which methods protect the underlying data well and which methods preserve the data structure.

method	uncorrelated		correlated		uncor, outl.		cor, outlier		cor, outlier	
	prot.	qual.	prot.	qual.	prot.	qual.	prot.	qual.	prot.	qual.
additive	pass	pass	pass	pass	part.	part.	part.	part.	pass	pass
correlated	pass	pass	pass	part.	pass	fail	pass	fail	pass	fail
correlated2	pass	pass	pass	pass	part.	part.	part.	part.	pass	pass
restr	pass	pass	pass	pass	pass	pass	pass	pass	pass	fail
ROMM	pass	pass	pass	pass	pass	pass	pass	pass	pass	fail
outdect	pass	pass	pass	pass	pass	pass	pass	pass	pass	pass
swappNum (p=15)	pass	pass	pass	fail	pass	fail	pass	fail	pass	part.
swappNum (p=40)	pass	pass	pass	fail	pass	fail	pass	fail	pass	fail
single	pass	fail	pass	fail	pass	fail	pass	fail	pass	part.
onedims	pass	pass	pass	pass	pass	part.	pass	part.	fail	pass
influence	pass	pass	pass	pass	pass	part.	pass	part.	pass	pass
pca	pass	pass	pass	pass	pass	part.	pass	pass	pass	fail
clustpppca	pass	pass	pass	pass	pass	pass	pass	pass	pass	pass
mdav	pass	pass	pass	pass	pass	pass	pass	pass	pass	pass
rmd	pass	pass	pass	pass	pass	pass	pass	pass	pass	pass
gadp	pass	pass	pass	pass	pass	pass	fail	pass	pass	fail
shuffle	pass	pass	pass	pass	fail	pass	fail	pass	pass	fail
robGadp	pass	pass	pass	pass	pass	pass	pass	pass	pass	pass
robShuffle	pass	pass	pass	pass	pass	pass	pass	pass	pass	pass

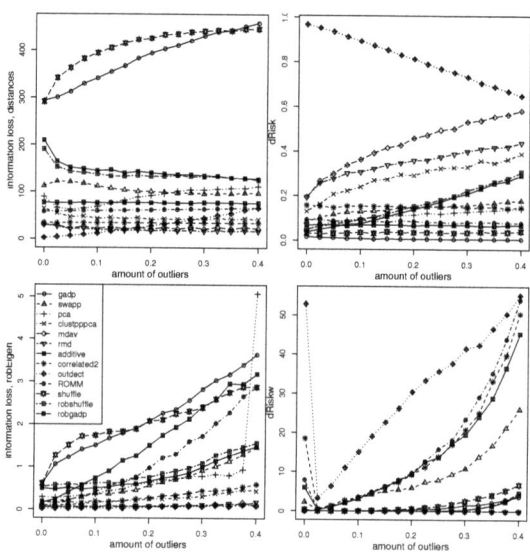

Figure 3.3: TOP LEFT: IL1s information loss measure versus the amount of shifted outliers. TOP RIGHT: disclosure risk (Mateo-Sanz et al. [2004b]) versus the amount of outliers in the generated data sets. BOTTOM LEFT: Information loss based on differences of the eigenvalues of the robust covariances between original and perturbed data versus the amount of outliers. BOTTOM RIGHT: weighted disclosure risk based on robust Mahalanobis distances versus the amount of outliers.

4 Robust Statistics Meets SDC: New Disclosure Risk Measures for Continuous Microdata Masking

[2]Published in "Privacy in Statistical Databases. Lecture Notes in Computer Science", Springer [Templ and Meindl, 2008a]

Matthias Templ[*,**], Bernhard Meindl[*],

* Department of Methodology, Statistics Austria, Guglgasse 13, 1110 Vienna, Austria. (bernhard.meindl@statistik.gv.at) and

** Department of Statistics and Probability Theory, Vienna University of Technology, Wiedner Hauptstr. 8-10, 1040 Vienna, Austria. (templ@statistik.tuwien.ac.at)

Abstract: The aim of this study is to evaluate the risk of re-identification related to distance-based disclosure risk measures for numerical variables. First, we overview different - already proposed - disclosure risk measures. Unfortunately, all these measures do not account for outliers. We assume that outliers must be protected more than observations near the center of the data cloud. Therefore, we propose a weighting scheme for each observation based on the concept of robust

Mahalanobis distances. We also consider the peculiarities of different protection methods and adapt our measures to be able to give realistic measures for each method. In order to test our proposed distance based disclosure risk measures we run a simulation study with different amounts of data contamination. The results of the simulation study shows the usefulness of the proposed measures and gives deeper insights into how the risk of quantitative data can be measured successfully. All the methods proposed and all the protection methods plus measures used in this paper are implemented in R-package *sdcMicro* which is freely available on the comprehensive R archive network (http://cran.r-project.org).

Keywords: Statistical disclosure control, Distance based disclosure risk, Outlier, Simulation study

4.1 Introduction

For many applications the measurement of disclosure risk is based on the idea of uniqueness, rareness, k-anonymity (Sweeney [2002]), base individual risk estimation (Benedetti and Franconi [1998], Franconi and Polettini [2004a]) or on certain models (Elamir and Skinner [2006]).

However, if the data consists of continuous scaled variables (e.g. business data on enterprises) other definitions of disclosure risk must be considered.

Applying the concept of uniqueness and k-anonymity on these quantitative variables results that every observation in the data set is unique.

If detailed information about a value of a numerical variable is available, one may be able to identify and eventually gain further information about an individual. So, an attacker may be able to identify statistical units by using for example linking procedures. The anonymization of numerical variables should avoid the successful merging of underlying data with other data sources.

We assume that an intruder has information about a statistical unit which is

CHAPTER 4. NEW DISCLOSURE RISK MEASURES

4.1. INTRODUCTION

included in the data and the intruder's information about some values of certain variables overlap with some variables in the data, i.e. we assume that the intruder's information can be merged with the data. In addition to that we assume that the intruder is sure that the link to the data is correct, except for microaggregated data. In this case the intruder can never be sure because at least k observations have the same value for each numerical variable.

In the next part of the introduction we will give a short overview of some popular distance based disclosure risk measures. In the next section we will describe the need of a special treatment of outliers that exist in almost every data set in Official Statistics. We then propose new measures of disclosure risk which give more realistic results when applied to data which include outliers. Finally, we compare all the measures considered in this study with a practical real data example as well as within a large simulation study.

All our proposed measures have been included in R-package *sdcMicro* (see e.g. in Templ [2007b], Templ [2008d]).

4.1.1 Distance Based Disclosure Risk Measures

By using distance based record linkage methods one tries to find the nearest neighbours between observations from two data sets. Domingo-Ferrer and Torra [2001] has shown that these methods outperform probabilistic methods. Such probabilistic methods are often based on the EM-algorithm which is highly influenced by outliers.

Another approach based on cluster analysis is provided by Bacher and Brand [2002] who uses k-means clustering with a high amount of clusters on mixed scaled variables. However, there are much better clustering methods available (see e.g. Templ et al. [2008]). k-means should not be applied on mixed scaled variables and, to put it crudely, this approach works as the usual distance based record linkage because it is based on the idea of similar objects and distance metrics.

Another type of measures of disclosure risk - referred to as value disclosure risk - is extensively used, e.g. by Muralidhar et al. [2006]. The main goal is to evaluate the gain in explanation of parameters or variables when releasing perturbed data.

Mateo-Sanz et al. [2004b] uses distance based record linkage and interval disclosure. In the first approach they search for the nearest neighbour from each observation of the masked data value to the original data points. Then they sign those observations for which the nearest neighbor is the corresponding original value. In their second approach they check if the original value falls within an interval centered on the masked value. Then they calculate the length of the intervals based on the standard deviation of the variable (method *SDID*).

In addition to that they define a rank-based interval procedure which is similar to the idea of *rank swapping* (method *RID*). For each variable of the masked data set they define a rank-based interval around each value. The rank-based interval includes p-percent of the total number of observations of the ranked variable. The proportion of the original values which fall into the calculated interval is used as measure of disclosure risk.

The calculation of an interval is based on a vector k of length p, the dimension of the confidential variables. k indicates how large these intervals for each variable are. In the implementation of *sdcMicro* the elements of k are set to 0.01 by default.

4.2 Special Treatment of Outliers for Disclosure Risk

Almost all data sets from Official Statistics consists of statistical units whose values in at least one variable are quite different from the main part of the observations. This leads to the fact that these variables are very asymmetric distributed. Such outliers might be enterprises with a very large value for turnover, for example, or persons with extremely high income or even multivariate outliers.

Unfortunately, an intruder may have a big interest in the disclosure of a large enterprise or of an enterprise which has specific characteristics. Since enterprises are often sampled with certainty or have a sampling weight near to 1 the intruder can be very confident that the enterprise he wants to disclose is definitely in the sample. In contrast to that an intruder may not be as interested to disclose statistical units which have the same behaviour than the main part of the observations. For these reasons it is reasonable to define measures of disclosure risk that take the "outlyingness" of an observation into account.

Therefore we assume that outliers should be much more perturbed than non-outliers because they are easier to re-identify even when the distance from the masked observation to its original observation is relatively large.

4.2.1 "Robustification" of *SDID*

In a first step we robustify method *SDID* because it is obvious that outliers increase the intervals estimated with method *SDID* dramatically since the calculation of the classical standard deviation is based on squared distances between the observations and the arithmetic mean.

However, method *SDID* can be easily robustified by using a robust measure for the standard deviation. We propose to use the MAD instead of the classical standard deviation. The MAD is given by

$$\text{MAD} = 1.4826 * median(|x_i - \tilde{x}|) \;,$$

with $\tilde{x}$ being the median and the constant $= 1.4826$ ensures consistency. We will call this robustified method *RSDID*.

4.3 New Measures of Disclosure Risk

All disclosure risk intervals obtained from methods *SDID*, *RID*, from the methods based on cluster analysis as well as *RSDID* do not depend on the scale of the actual value and therefore, the length of the interval is equal for non-outlying and outlying values. Thus, we now propose new, more realistic measures of disclosure risk which account for the "outlyingness" of each observation.

Mateo-Sanz et al. [2004b] searches for outliers (after z-transformation of the variables) by calculating Euclidean distances for each observation with respect to the origin. Finally the observations are sorted based on these distances and the five% farthest observations are classified as outliers. Nevertheless, this approach is quite poor for the detection of outliers because of using a non-robust transformation as well as using Euclidean distances in a multivariate space.

The aim is now to measure the distance of each observation to the center of the data in a multivariate space. For a p-dimensional multivariate sample x_i ($i = 1, \ldots, n$) the Mahalanobis distance is defined as

$$\mathrm{MD}_i = (x_i - t)^T C^{-1}(x_i - t) \quad \text{for} \quad i = 1, \ldots, n \;, \tag{4.1}$$

where t is the estimated multivariate location and C the estimated covariance matrix. Usually, t is the multivariate arithmetic mean, and C is the sample covariance matrix.

Multivariate outliers may simply be defined as observations featuring large (squared) Mahalanobis distances. However, this approach has several shortcomings which are visualized in Figure 4.1. The concept of classical Mahalanobis distances fails completely in this example and does not describe the behaviour neither of the outliers nor of the homogeneous part of the data well. Single extreme observations as well as groups of observations that depart from the main data structure can have a severe influence on this distance measure because both location and covariance are usually estimated in a non-robust manner.

The fast minimum covariance determinant (MCD) estimator (Rousseeuw P.J. [1999]) is well known in the literature and has been used to estimate the location and the covariance structure in a robust way. Using this estimator in formula

CHAPTER 4. NEW DISCLOSURE RISK MEASURES
4.3. NEW MEASURES OF DISCLOSURE RISK

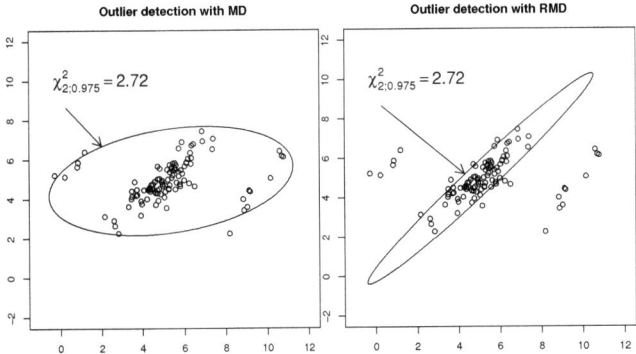

Figure 4.1: Illustration of the concept of Mahalanobis distances and robust Mahalanobis distances on a simply 2-dimensional example. LEFT: Tolerance ellipse (95 %) and "outlier detection" using Mahalanobis distances. RIGHT: Tollerance ellipse (95 %) and outlier detection using robust Mahalanobis distances.

4.1 leads to robust Mahalanobis distances (RMD). Figure 4.1 shows that Mahalanobis distances based on classical measures are not suited for the definition of disclosure risk intervals and that the robust version needs to be chosen. Observations whose RMD_i is greater than $\chi^2_{(0.975,p)}$ may be defined as outliers. This is however only an approximation since squared RMD are only approximately χ^2 distributed (see e.g. in Filzmoser [2004]).

The intervals for each data value should now depend on the robust distances, i.e. the intervals may be defined as $k_j \times (RMD_i)^{1/2}$, $j \in \{1, \ldots, p\}$. Following this approach we obtain a disclosure risk for each observation by checking if any value of an observation falls into the corresponding interval or not. We then calculate the percentage of observations featuring high risk and call this procedure *RMDID1*.

Since we want to consider the "outlyingness" of each observation we simply weight each observation with its RMD or with $(\text{RMD}_i)^{1/2} \cdot k_j$. In this paper we use the latter weight and call this new procedure *RMDID1w*.

The need for an additional approach can simply be met by using a microaggregation procedure for the perturbation of microdata. We assume that we have applied microaggregation with high aggregation level, e.g. 10. All the methods described previously provide a high risk of disclosure if the original value and the microaggregated value are close to each other. But these measures are unrealistic for this simple microaggregation example since 10 observations possess the same value in the microaggregated variable, and data intruder can never be sure which one is the correct link. Especially, if this observation is near the center of the data cloud the previous measures fail to provide a meaningful measure of disclosure risk.

These problems are solved by looking closely at observation that have a relatively high risk of re-identification in *RMDID1* or *RMDID1w*. An observation which is marked as unsafe (with method *RMDID1* or *RMDID1w*) is considered safe if m observations are very close to the masked observation (we call this procedure *RMDID2*). This problem is illustrated in Figure 4.7 with a simple 2-dimensional example data set that is described below.

We now describe the proposed algorithm as follows:

1. Robust Mahalanobis distances are estimated in order to get a robust multivariate distance for each observation.

2. Intervals are estimated for each observation around every data point of the original data points where the length of the intervals are defined/weighted by squared robust Mahalanobis distances and the parameter k_j. The higher the RMD of an observation the larger the corresponding intervals.

3. Check if the corresponding masked values fall into the intervals around the original values or not. If the value of the corresponding observation lies within such an interval the entire observation is considered unsafe. We obtain a vector indicating which observations are safe or not ($\to$ we are finished already when using method *RMDID1*).

4. For method *RMDID1w* we calculate the weighted (using RMD) vector of disclosure risk.

5. For method *RMDID2*: whenever an observation is considered unsafe we check if m other observations from the masked data are very close (defined by a parameter $k2$ for the length of the intervals as for *SDID* or *RSDID*) to this observation using Euclidean distances. If more than m points are within these small intervals we conclude that the observation is "safe".

For measures *SDID* and *RSDID* the parameter vector k is a multiplier of the standard deviation. For methods *RMDID1* and *RMDID2* k is a multiplier of the squared RMD. While for standardized data sets the standard deviation is one and the interval around the masked value x_i^{mask} has a length of $2 \cdot k_i$ the RMD weights this interval according to $(\text{RMD}_i)^{1/2}$.

Naturally, most of the intervals corresponding to values in the center of the data are down weighted, and only for those observations which are away of the center of the data cloud the intervals increase.

The second parameter vector $k2$ for method *RMDID2* which evaluates if the masked data has any close neighbours can be set at, for example, 0.05 for each $i \in 1, \ldots, p$. We look for (automatically) standardized data whose values are within an interval of length $2 \cdot k2_i$ around $x_{i(mask)}$.

Figure 4.2 points out the idea of weighting the disclosure risk intervals. While for method *SDID* and *RSDID* the rectangular regions around each value are the

CHAPTER 4. NEW DISCLOSURE RISK MEASURES
4.3. NEW MEASURES OF DISCLOSURE RISK

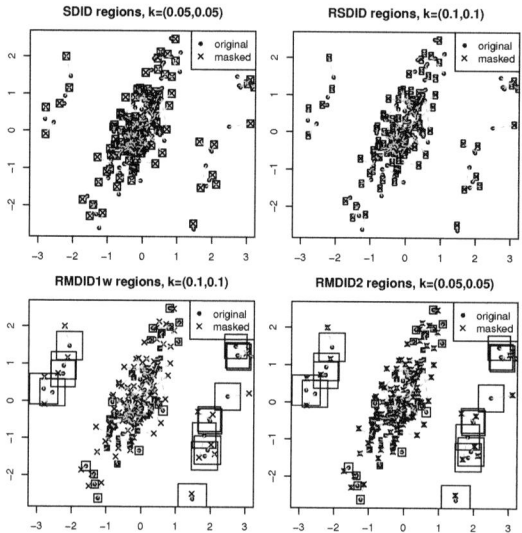

Figure 4.2: Original observations and the corresponding masked observations (perturbed by adding additive noise). In the bottom right graphic small additional regions are plotted around the masked values for RMDID2 procedure.

same as for each observation our proposed methods take the RMD of each observation into account. The difference between the bottom right and the bottom left graphic is that for *RMDID2* rectangular regions around each masked variable are calculated as well. If an observation of the masked variable falls into an interval around the original value it is checked if this observation does have close neighbours, i.e. if the values of m other masked observations are inside a second interval around this masked observation.

While it is not possible to interpret the weighted disclosure risk measures *RM-*

DID1w and RMDID2 in a probabilisitic way. However, the proportion of unsafe values on all observations using the unweighted measures can be interpreted as a global, probabilistic measure of disclosure risk.

4.4 An Example Using the Tarragona Data Set

Please note, when applying the related functions in *sdcMicro* no data standardisation needs to be done since both the center and the scatter of each variable are already considered in our implementation. This means that the standardisation is done automatically.

The first result is obtained by using microaggregation method *rmd* from package *sdcMicro*. This algorithm is called RMDM (**R**obust **M**ahalanobis **D**istance based **M**icroaggregation) which was proposed by Templ [2007b] and has excellent properties (see e.g. the results of the simulation study in Templ and Meindl [2008b]). We are interested in how the parameter vector k influences the results of the disclosure risk measures. In this example which results in Figure 4.3 we set k as a vector as large as the dimension of the Tarragona data set. k varied - with equal values - from 0.01 to 0.2.

One can see that all of the disclosure risk measures account for the increase of the interval length. Naturally, measure *RMDID2* is always zero because microaggregation of the data was conducted and a search for near neighbours was done. Of course, if the parameter m is increased then this measure is quite similar to *RMDID1w*.

4.5 Simulation Results

Based on the proposed measures the disclosure risk is evaluated in a simulation study with 1000 simulations similar to the approach in Templ and Meindl [2008b]. We generate 1000 data sets of dimension 300×2 that are multivari-

CHAPTER 4. NEW DISCLOSURE RISK MEASURES
4.5. SIMULATION RESULTS

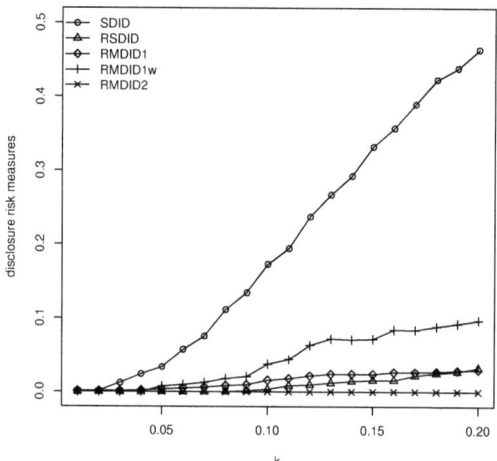

Figure 4.3: Influence of parameter k on different distance-based disclosure risk measures.

ate normal distributed with mean vector $\mu = (0,0)$ and covariance matrix Σ ($\sigma_{ii} = (1,1), \sigma_{ij} = (0.8, 0.8)$). Furthermore, we include a shifted outlier group with mean $\mu_{out} = c(10,0)$ and the same variance-covariance structure and calculate all risk measures discussed above for any of the generated data sets.

Please note that we only show medians of the simulation results in order to stay within the limit of pages.

Figure 4.4 shows the influence of outliers on the *SDID* disclosure risk measure and the deviation of the eigenvalues from the robust estimated covariance matrix between the original and the masked data. However, other measures can be used too (see Mateo-Sanz et al. [2005]). This measure of information loss was chosen because the eigenvalues of the covariance matrix may be used to

CHAPTER 4. NEW DISCLOSURE RISK MEASURES

4.5. SIMULATION RESULTS

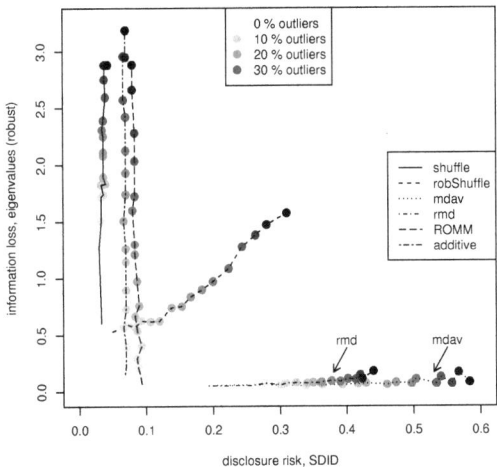

Figure 4.4: Effects of shifted outliers (0 till 40 percent in 2.5 percent steps) on some protection methods based on *SDID* measure.

represent the data structure and are input of popular multivariate techniques like (robust) principal component analysis. The robust estimation of the covariance matrix is done using the fast MCD-estimator (Rousseeuw P.J. [1999]). A robust estimation may be prefered since traditional measures may give unrealistic results on inhomogeneous data sets.

It is easy to see that the disclosure risk does not increase when using method *shuffling*. Templ and Meindl [2008b] showed that this relates to the meaningless results of *shuffling* when data feature outliers. Increasing the number of outliers in the data results in high amount of information loss also for method *additive noise addition* and *ROMM* (see the description of these methods in Muralidhar and Sarathy [2006], Brand and Giessing [2002] and Ting et al. [2005]). *Robust shuffling* (Templ and Meindl [2008b]) and especially *mdav* (see e.g. Hundepool et al. [2007b]) perform better and method *rmd* (Templ [2007a]) outperforms all the methods.

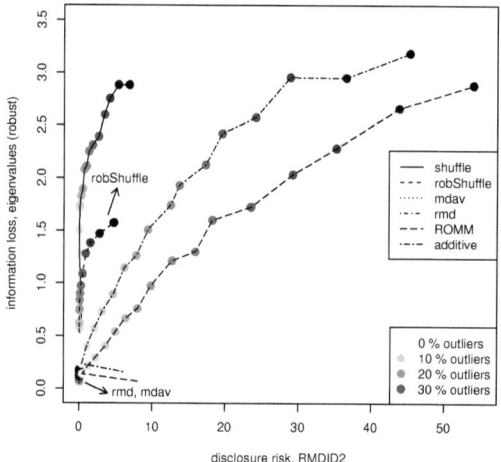

Figure 4.5: Effects of outliers on some protection methods based on *RMDID2* measure.

While the information loss for methods *additive*, *shuffling* and *ROMM* is comparable in Figure 4.5, the disclosure risk for these methods is relatively low compared to the *shuffling* procedure. But again, all these methods are strongly influenced by outliers and provide worst results regarding information loss. *Robust shuffling* is again performing much better than *shuffling* if outliers are included in the data. Again, *rmd* and *mdav* perform best. The "shift" from zero outliers to 2.5 percent outliers that is visible for some methods can be explained because the disclosure risk decreases when the perturbation goes down as soon as outliers appear in the data.

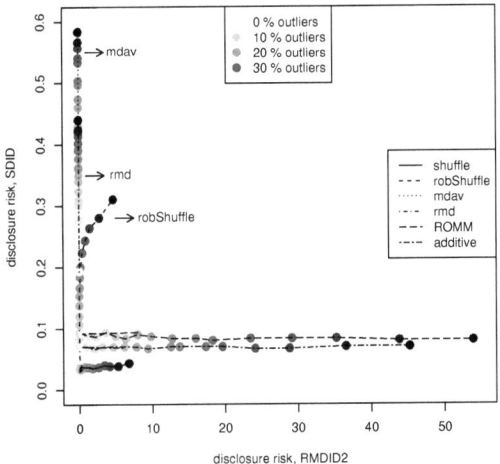

Figure 4.6: Comparison of *SDID* and *RMDID2* measure under different contaminations.

In Figure (4.6) the *SDID* distance based disclosure risk measure is plotted versus the *RMDID2* measure. It is clearly visible that *SDID* fails and does not account for outliers, i.e. does have the same length of disclosure risk intervals both for non-outliers and outliers. The disclosure risk does not increase, for example, for method *shuffling* since the data structure from the perturbed data is very different from the original one because of the non-robustness of *shuffling* (see also in Templ and Meindl [2008b]). In contrast to the other risk measures, *RMDID2* considers the observations as safe when using microaggregation and parameter $m < g$, with g being the aggregation level.

We also want to find out if outlying observations do have a high risk of disclosure. Thus, we divided the data in outliers and non-outliers and visualize the results for the outlier part for every single outlying observation.

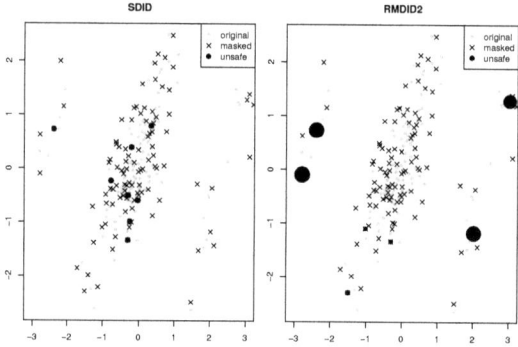

Figure 4.7: Disclosure risk evaluated for every observation.

Since we evaluate the disclosure risk which is weighted by the RMD for every observation we can simply evaluate which observation possess a high risk of re-identification. For the previous 2-dimensional example we can show which observations are considered unsafe (see Figure 4.7). The left graphic of Figure 4.7 shows the unsafe observations discovered by method *SDID*. One can see that some outliers are not considered unsafe although the masked observations is relatively near to the original one. *RMDID2* accounts for this and considers some outliers as unsafe as can be seen on the right area of Figure 4.7. In addition to that each observation does have a different risk. Therefore, one can easily perform additional protection on these observations (e.g. adding noise) as long as the newly proposed measure outlay a risk higher zero or higher a predefined threshold.

4.6 Conclusion

When considering that data includes outliers (which is the case for virtually any real life data set) we have to tackle two problems considering statistical disclosure control for numerical variables. The first problem is that outliers may disturb the protection methods (or make the generation of adequate synthetic data impossible) with a high loss of information. This problem was partly considered in this study and a deeper insight into this problem was given by Templ and Meindl [2008b]. The second problem - the problem which is mainly discussed in this paper - is that outliers must be protected more than the observations which are located near the center of the data cloud, i.e. which are having low robust Mahalanobis distances. We proposed new measures of disclosure risk called *RMDID1*, *RMDID1w* and *RMDID2* that account for these problems and that assign a risk to every observation weighted by the robust Mahalanobis distance of the observation (*RMDID1w* and *RMDID2*). In addition to that we described the problem of microaggregation (but this is also related to all other methods) where an intruder can never be sure which of the aggregated values correspond to the original ones.

The new measures of disclosure risk provide realistic estimations on the risk of re-identification of each observation separately. Therefore, additional protection for high-risk observations may be provided to the masked data resulting in "protected" data with good quality with respect to both high data utility and very low disclosure risk.

All the methods proposed in this paper are freely available on the web and are included in R-package *sdcMicro*.

5 Software Development for SDC in R

Published in "Privacy in Statistical Databases. Lecture Notes in Computer Science" [Templ, 2006a]

Matthias Templ[*,**], Bernhard Meindl[*],

[*] Department of Methodology, Statistics Austria, Guglgasse 13, 1110 Vienna, Austria. (bernhard.meindl@statistik.gv.at) and

[**] Department of Statistics and Probability Theory, Vienna University of Technology, Wiedner Hauptstr. 8-10, 1040 Vienna, Austria. (templ@statistik.tuwien.ac.at)

Abstract: The production of scientific-use files from economic microdata is a major problem. Many common methods change the data in a way which leaves the univariate distribution of each of the variables almost unchanged towards the distribution of the variables of the original data, the multivariate structure of the data, however, is often ruined.

Which methods are suitable strongly depends on the underlying data. A program system with which one can apply different methods and evaluate and compare results from different algorithms in a flexible way is needed. The use of methods for protecting microdata as an exploratory data analysis tool requires a powerful program system, able to present the results in a number of easy to grasp graphics. For this purpose some of the most popular procedures for anonymising micro data are applied in a flexible R-package. The R system supports flexible data import/export facilities and advanced developement tools for the development of such a software for disclosure control.

Additionally to existing algorithms in other software (MDAV algorithm for microaggregation, ...) some new algorithms for anonymising microdata are implemented, e.g. a fast algorithm for microaggregation with a projection pursuit approach. This algorithm outperforms existing other algorithms for most of real data.

For all this algorithms/methods print, summary and plot methods and methods for validation are implemented.

In the field of economics suppression of cells in marginal tables is likely to be the most popular method to protect tables for statistical agencies. The use of linear programming for cell suppression seems to be the best way of protecting tables and hierarchical tables.

Some R-packages for various fields of disclosure control are being developed at the moment. It is easy to learn the applications of disclosure control even with little previous knowledge because of its integrated online-help with examples ready to be executed.

5.1 Using R for Disclosure Control

R (R Development Core Team [2008a]) is a open source statistics software package subjected to the GPL and therefore free and extendable for companies. R can be downloaded from the following website:

$$\text{http://cran.r-project.org/}$$

CHAPTER 5. SOFTWARE DEVELOPMENT FOR SDC IN R
5.1. USING R FOR DISCLOSURE CONTROL

Many methods and papers have been presented on disclosure control over the last years, but the underlying code is rarely available. The implementation of methods in software is required to evaluate the quality of the methods. For this purpose the free available, open-source, object-oriented, high-level language software R seems to be perfect.

Nowadays, R has become the standard statistical software. Thousands of people are involved in the development of R both at universities and companies and more than 700 add-on packages have been built in the last years.

R can deal with many different data formats and have very flexible data import/export facilities which is quite important when dealing with data from various formats. R can also communicate with various popular software and data bases. A major advantage of using R for disclosure control is that the facilities of R and also already implemented algorithms and graphical tools can be used easily. We do not need develop things new.

Applying different methods for disclosure control on data and evaluate and compare the results is a kind of explorative data analysis. For this purpose a object-oriented language like R is quite important.

Also very important is the reproducibility of results (Leisch and Rossini [2003]) when applying algorithms for disclosure control on data and when doing a validation of the results or comparing different results and making some nice graphics. For this, R is very well designed and in combination with LaTeX someone can produce dynamical reports, where LaTeX-code and R-code can be written and executed in/from one document together with the help of Sweave (Leisch [2002b], Leisch [2002a]).

Additionally it is very easy to develop your own packages with online help-files in R. With some developement tools of R one can make nice graphical user interfaces (GUI) like the Argus GUI's (Hundepool et al. [2003]) as well.

Everybody can contribute code to the packages and help to make the project a success. The code is freely available under GPL and legally protected for commercial use of others (nobody has the right to use the code for an commercial implementation in software). The open source status and free use of the code should help to make the code better and better. Everybody is invited to check and upgrade the code.

The R system provides a powerful programming language and existing Fortran or C-code can easily embedded. R is the most powerful program system in the statistical world and in my opinion we don't have an alternative to R in the near future.

5.2 Microdata protection

The methods for making microdata confidential vary considerably, depending on the scaling of the data. We will concetrate on continuous microdata only in order to stay within the limit of pages of this paper.

We want to give data to researchers and preserve confidentiality at the same time. There are few general concepts to do this. Some statistical agencies have decided to design confidentiality preserving model servers. E.g. the United States Bureau of the Census operates a number of *Research Data Centers* where researchers with special sworn status have access to specified microdata (see e.g. Steel and Reznek [2005]). Researchers can apply models on data which can not be seen and the results of the models are checked by the Census staff. This approach preserves confidentiality but is not flexible, and in my point of view not compatible with a modern statistical world. When working with model servers the result of the model is the object of interest not the underlying data. But it is really difficult to apply methods and it is problematic to choose and evaluate models without seeing the data. The model can e.g. be influenced by outliers. Additionally, only very few methods are available on such model servers.

Much more flexible are remote access facilities. Researches can look at the data and can choose a suitable method for analysing the underlying data. Finally, the output should be checked by the staff of the statistical agencies and is, confidentiality preserved, usually to sent per e-mail the researchers (see e.g. Hundepool and de Wolf [2005] or Borchsenius [2005]).

It is expensive and time-consuming to implement one of these two concepts.

The third approach for preserving confidentiality is to produce scientific use files by perturbation of microdata. The main goal is to produce a data set from the original data which preserves confidentiality and has a same structure as similar to the original data set as possible.

There are some concepts for this approach. The well-established concepts are *Microaggregation* (Anwar [1993]), *Adding Noise* (see e.g. Kim [1986], Kim and Winkler [1995]), *Rank Swapping* (Dalenius and Reiss [1982]), *Blanking and Imputation* (Griffin et al. [1989]) and the generation of *synthetic data* with the same stucture as the original data (e.g. *Latin Hypercube Sampling* Iman and Conover [1982], Stein [1987], Wyss and Jorgensen [1998]).

All these methods have been implemented in various R-Packages.

5.3 Microaggregation

On *htttp://neon.vb.cbs.nl/casc/Glossary.htm* we can find the "official" definition of Microaggregation: *"Records are grouped based on a proximity measure of variables of interest, and the same small groups of records are used in calculating aggregates for those variables. The aggregates are released instead of the individual record values."* (Elliot et al. [2005]). More references can be found in (Anwar [1993], Defays and Nanopoulos [1993], Defays and M.N. [1998], Domingo-Ferrer and Mateo-Sanz [2002])

While for the proximity measure very different concepts can be used, microaggregation is naturally done with the mean. Nevertheless, other measures of location can be used for aggregation, especially when the group size for aggregation has been taken higher than 3. Since the median seems to be unsuitable for microaggregation due to it's rather high *breakdown point*, other measures like an *onestep from median* (see e.g. Hulliger [1999]) can be chosen. The *breakdown point* of an estimator measures the maximal percentage of the data points that may be contaminated before the estimates becomes completely corrupted.

5.3.1 Clustering at First

The package contains also a method with which the data can be clustered with a variety of different clustering algorithms. Clustering the observations before applying microaggregation might be useful. There is quite a lot of algorihms to do a clustering with, but for most of these cases Mclust (Fraley and Raftery [1998]) provides the best results. These technique, which is not based on distance measures, usually find the clusters by optimizing a maximum likelihood function. Avoid using hierarchical or classical partitioning cluster algorithms because hierarchical clustering algorithms result in worst partitions and classical partitioning algorithms, such as kmeans, result in spherical clusters with nearly the same size. Note that in our approach the data should be standardised before clustering, especially when the variables are of unequal scaling. Without standardising the data one variable might have the highest influence in each cluster and this is not what we want. Cluster analysis in general does not need normally distributed data. However, it is advisable that heavily skewed data are first transformed to a more symmetric distribution. If a good cluster structure exists for a variable, we can expect a distribution which has two or more modes. A transformation to more symmetry will preserve the modes but remove large skewness.

5.3.2 Methods Based on Sorting of Variables

We have developed a package called Microaggregation which contains methods like *individual ranking* (Defays and M.N. [1998]), sorting based on a single variable Defays and Nanopoulos [1993], Anwar [1993] and some related methods.
Cluster analysis can be applied before applying these methods. The clustered data can be sorted in each cluster depending on the most important variable in each of the clusters (in the package we call this method *influence*).

5.3.3 Projection Methods and MDAV

Package Microaggregation contains a good method called *mdav* **M**aximum **D**istance to **A**verage **V**ector (*mdav* is in turn an evolution of the multivariate fixed-size microaggregation in Domingo-Ferrer and Mateo-Sanz [2002] proposed by the same authors). This method was first implemented in the μ-Argus software (Hundepool et al. [2005]).

Another approach is to sort the data according to the first principal component (see e.g. in Schmid [2006]) which is a well-documented method in SDC. Classical Principal Component Analysis (PCA) (Pearson [1901]) is very sensitive to outlying observations since it is computed from eigenvalues and eigenvectors of the non-robust sample covariance matrix. Therefore applying PCA to sort the observations on the first principal component before aggregation may provide worst results in context of microaggregation.

In addition to that, package Microaggregation contains two major types for a robustification of this approach.
The first one calculates eigenvectors and eigenvalues based on robust estimates of the covariance matrix. The MCD-estimation (Rousseeuw [1985]) is the default for the estimation of the covariance matrix. Others, like M-estimation (Maronna [1976]), the MVE estimator (Rousseeuw [1985]), the orthogonalized Gnanadesikan-

Kettering (OGK) estimator (Maronna and Zamar [2002]) and some more can also be used. High breakdown point estimators for the covariance matrix are to be prefered. When using classical PCA or these robustified PCA all principal components must be estimated, but in the context of microaggregation we need only the first principal component.

The second approach avoids this and estimates the first (robust) principal component without covariance estimation. Croux and Ruiz-Gazen [2005] has developed a method based on *projection pursuit* (PP) Li and Chen [1985], Huber [1985]. With PP we search for directions with maximal variance of the data projected on it. Instead of using the classical variance estimator they use a robust scale estimator S_n as *projection pursuit index*. For a sequence of observations $x_1, \ldots, x_n \in \mathbb{R}^p$, the first "eigenvector" is defined as

$$v_{S_n,1} = \underset{||a||=1}{\mathrm{argmax}} S_n(a^t x_1, \ldots, a^t x_n) . \tag{5.1}$$

The associated "eigenvalue" is then, by definition,

$\lambda_{S_n,1} = S_n^2((v_{S_n,1})^t x_1, \ldots, (v_{S_n,1})^t x_n).$

Li and Chen proposed working with an M-estimator of scale for S_n, and applied a general projection-pursuit algorithm for maximizing Formula (5.1), leading to an iterative and complicated computer intensive method. Nowadays there is a renewed interest in the projection-pursuit approach to PCA. Filzmoser Filzmoser [1999] for instance applied it to a geostatistical problem. Perhaps the best known robust dispersion measure is the Median Absolute Deviation (MAD). For a sample $\{x_1, \ldots, x_n\} \subset \mathbb{R}$ it is defined as

$$\mathrm{MAD}_n(x_1, \ldots, x_n) = 1.486 \, \underset{i}{\mathrm{med}} |x_i - \underset{j}{\mathrm{med}} x_j| , \tag{5.2}$$

where the constant 1.486 ensures consistency at normal distributions.

Primarily when having mixed structures in your data it is a good idea to cluster the data and apply the projection methods on each cluster. This can be easily done by setting up an optional parameter in a function in package `Microaggregation`.

It is really easy to use the package Microaggregation and apply its methods, because of the online-help files, the included examples and the simple handling of objects in R.

5.4 Adding Noise

5.4.1 S4-class style

A S4-class R-Packages called AddNoise was developed. Normally in other statistical softwares you have only few classes, like *numeric*, *character*, or *data set*. In R there are much more classes and additionally one can design your own classes (for e.g. class *addNoise*). The concept of S4-classes (Chambers [1998]) in R is new and an extension of the traditional S3-class system in R (Chambers [1998]). Only few packages are written in S4 style up to now. S4 style code is very formal and you can define classes, plot-, print-, and summary methods. The advantages for the user of an S4-class package are that the packages are really flexible in use and mostly easy extendable with your own code. Additionally the user gets very precise error messages when operating the implemented functions in an incorrect way.

5.4.2 Methods

Beside the implementation of simple adding normal distributed noise and correlated noise (Kim [1986]), there are an implementation of Random Orthogonal Matrix Masking, called ROMM (Ting et al. [2005]). Note that this procedure preserve no confidentiality, e.g. the output of *biplots* (Gabriel [1971]) from the masked data and the original data are the same.

Additional there is a another concept of adding noise. Presume that observations which can be identified with diagnostic plots are confidential, then we can detect such observations with robust outlier detection methods. So, only these observations should be perturbed (and of course observations which can be identified with the help of key variables as well), depending on the outlyingness of these observations.

For the detection of critical observations, which may be identified by an data intruder, there are some (robust) outlier detection tools, like mahalanobis distances, robust mahahlanobis distances (see e.g. in Maronna [1976]), jackknifing for results on the first 2 eigenvalues (Efron and Tibshirani [1993]) and also on some univariate statistics which can be applied on Box-Cox transformed data (Box and Cox [1964]). A very good R-package for outlier identification is also package mvoutlier from Filzmoser Filmoser [2004].

5.5 Other approaches for continuous microdata

A package called rankSwapp (*rank swapping*) and a few methods, like *latin hypercube sampling* and *blanking and imputation* have been developed.

While with the rank swapping approach the univariate structures of the data are nearly the same as for the original data, the multivariate structure is modified dramatically.

The results of latin hypercube sampling are not satisfactory, even not when doing some iterations.

5.6 Validation of the Results from Microdata Protection

First we want to give a short overview about existing validity measures, which are nearly almost univariate measures of information loss. After these, we want to propose other measures of information loss, which evaluates the multivariate structure of the original and the perturbed data.

One measure of information loss which is proposed by Mateo-Sanz et al. [2004b]

is given by the original and the perturbed version of a observation i

$$IL1 = \frac{1}{d} \sum_{i=1}^{p} \frac{|x_{ij} - x'_{ij}|}{\sqrt{2}S_j} \qquad (5.3)$$

where S_j is the standard deviation of the j-th variable in the original data set. This measure of information loss does not evaluate how well univariate or multivariate statistics are preserved. This is a real disadvantage of this kind of measures and we want to show other measures which takes univariate and multivariate statistics into account.

In Mateo-Sanz et al. [2004b] there is also proposed a measure of disclosure risk, which based on distances and assumes that an intruder has additional information (disclosure scenarios) so that one can link the masked record of an individual to its original version Mateo-Sanz et al. [2004b]. Given the value of a masked variable, they check whether the corresponding original value falls within an interval centered on the masked value. The width of the interval is based on the rank of the variable or on its standard deviation Mateo-Sanz et al. [2004b].

Applying these measures on real data from a subset of the structural business statistics in Austria (one economic sector and 5 variables) we will see the IL1 and disclosure risk on the following graphics resulting from a part of the implemented algorithms for disclosure control. The algorithms which are chosen are M3 - M11 (microaggregation with aggregation level from 3 to 11 on different methods, ROMM with different parameters for the magnitude of perturbation Ting et al. [2005], rank swapping with different maximum rank differences, which is expressed as a percentage of the total number of records (1%, 5%, 10%, 20%) and latin hypercube sampling Iman and Conover [1982].

Previous Figures shows the polarity from data utility (Figure 1) and disclosure risk (Figure 2). Unfortunately, the previous measure of information loss tells us nothing about the quality of the perturbation, but we can anticipated that our proposed method with the projection pursuit approach (clustpppca) e.g. is more

CHAPTER 5. SOFTWARE DEVELOPMENT FOR SDC IN R
5.6. VALIDATION OF THE RESULTS FROM MICRODATA PROTECTION

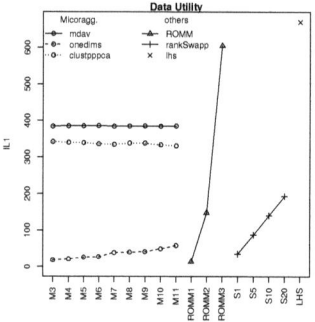

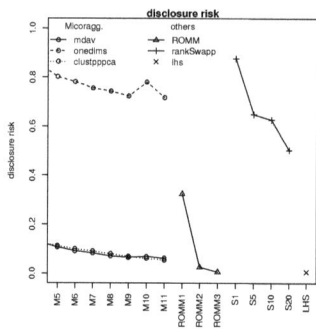

Figure 5.1: Data utility resulting from different perturbation methods

Figure 5.2: Disclosure risk resulting from different perturbation methods

suitable than method `mdav`. Note, that we want to have anonymised data, which have the same statistical properties as the original data.

In the following we will evaluate some methods on real data from the structural business statistics in Austria on some statistical measures of information loss. The results based only on microaggregation methods, because our experience shows that only microaggregation fulfills the required amount of anonymisation and we want to be in the limit of pages in the paper.

```
> method <- c("simple", "single", "onedims", "pca", "pppca", "influence",
+   "clustpppca", "mdav")
> g <- valTable(x = x, method = method, measure = "onestep")
> g[, -c(9, 12, 13, 14)]
\end{Sinput}
\begin{Soutput}
     method amean amedian aonestep devvar   amad   acov   acor   adlm apcaload
1    simple 1.809   0.538    0.186  2.859  0.582  1.430  0.606  0.012    0.099
2    single 0.995   0.322    0.301  2.933  0.318  1.466  2.573  0.017    0.051
3   onedims 0.100   0.007    0.004 66.549  0.007 33.274 39.750  0.194    1.416
4       pca 0.782   0.289    0.225  1.676  0.239  0.838  0.285  0.035    0.101
5     pppca 1.064   0.363    0.293  1.919  0.322  0.959  0.894  0.089    0.236
6 influence 0.943   0.207    0.231  2.169  0.289  1.084  1.107  0.041    0.222
```

7	clustpppca	0.996	0.226	0.161	2.025	0.229	1.013	1.577	0.057	0.369
8	mdav	1.225	0.267	0.226	3.131	0.364	1.566	7.022	0.016	0.288

The three columns of the previous table represents univariate measures of information loss, the following four columns represent multivariate measures of information loss followed by one columns representing differences in results from a classical regression models and followed by a mean difference of the loadings getting from a classical principal component analysis. A detailed description can be found in the online-help files from package Microaggregation. You can easily see that the individual ranking method (onedims) preserves univariate statistics quite well, but fails completely in the multivariate case. In many cases the principal component analysis method via projection pursuit on each cluster (found with a model based clustering algorithm) performs best.

Note that the data must not be aggregated with the mean. Sometimes it is useful to aggregate it with an another measure of location, like *onestep from median*. Observations, which are outside med$(x) \pm c$.mad(x) have been put to these limits and then the mean is calculated. c is a constant to be chosen as in Formula (5.2). Additionally other robust measures like M-estimates Huber [1985] can be used, where data outside an robust interval are down weighted.

To evaluate differences between the original data and the perturbed data (or between two perturbation methods) in the univariate and multivariate structure of these data a variety of univariate and multivariate comparison plots is implemented. One plot is e.g. to compare the covariance structure of the original data. In Figure 5.3 you can see the covariances of the original data (black lines) in comparison with the covariances of the perturbed data (blue lines). Here the perturbation is done with the pca microaggregation method. The robustified pca with projection pursuit in each cluster are shown in Figure 5.4. While the results with sorting based on first classical principal component looking not really good, the perturbed data resulting from the robustified version of pca looks very well.

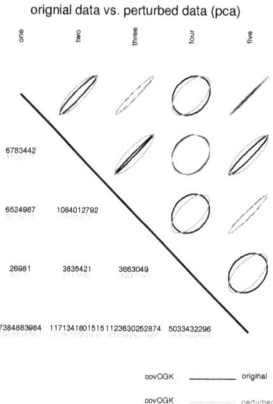
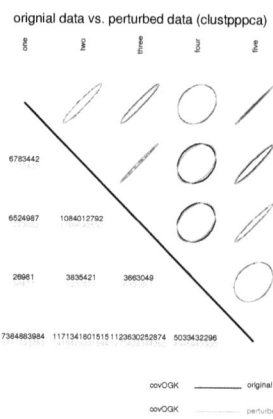

Figure 5.3: Comparison plot of original and microaggregated data via sorting on the first classical principal component.

Figure 5.4: Comparison plot of original and microaggregated data via sorting on the first robust principal component with projection pursuit in each cluster.

5.7 Protection of hierarchical tables

For this purpose a package called `disclosure` is under development.

The ordinary *primary cell suppression*, like *dominance rules*, p-percent rule, pq-percent rule (see e.g. in Elliot et al. [2005]) are implemented and additional cells can be easily suppressed by the user.

For *secondary cell suppression* there is a method called *disc* (Piker [1995]) implemented. It is similar to the Hypercube method (Repsilber [1994]). The disadvantage of this method is that after secondary cell suppression some primary protected cells can be computed too accurately and there is sometimes a little bit of over-suppression. A better strategy is to do a secondary cell suppression based on *linear programming*. The aim is e.g. to minimise the amount of suppressed cells (or other approaches) in (hierarchical) tables under the constraint that each primary suppressed cell can be computed only on a predefined interval (can not be computed too accurately). In R there is a wrapper function in C for the freely available (under LGPL2) linear program solver `lpSolve` (Berkelaar et al. [2006]) included, which can solve general linear/integer problems and more. Perturbation and controlled rounding methods are not implemented yet (you can do this with τ-Argus (Hundepool et al. [2003])).

The usage of this package should be very easy, but note that there is no general solution yet been implemented for applying the tools of package *disclosure* automatically on varying hierarchical structures. Even for intruding some tables the package can be used in a very simple way. After loading data this should be carried out in one statement with some optional arguments. The functionality should be seen clearly from the online-help files of the package.

In the following I will show only a attacker problem from data aggregated at European Union level. Here is an example of such an table which is already hidden from member states and rules for suppression from Eurostat:

5.7. PROTECTION OF HIERARCHICAL TABLES

```
[1] "Table 1: EU level"

  C1  C2  C3  C4  C5  C6   EU
1  20  70  90  30  20  30  260
2 500   1   1  50   1  70  623
3  10   3   6  25  50  20  114
4  70  80  10 100  30  40  330
5 600 154 107 205 101 160 1327
```

```
[1] "Table2: Supp. from member
states + rules for agg."

   C1  C2  C3  C4  C5  C6   EU
1  20  70  90  30  NA  30   NA
2 500  NA  NA  NA  NA  70  623
3  10   3   6  NA  NA  20  114
4  70  NA  NA 100  30  40   NA
5 600 154 107 205 101 160 1327
```

We can easily attack these table with our implemented functions based on linear programming: In the following the table on the left side shows us the attacker solution from Table 2. On the right side we can see the solution when all additional EU-aggregates have been hidden (note that only aggregates can be hidden and cells from member states must not be hidden).

```
> library(disclosure)
> i ← c(1, 2, 4, 5)
> lp.hier(e, e1)$lp.out2[, i]
       min max nrow ncol
 [1,]   18  71    1    5
 [2,]  258 311    1    7
 [3,]    0  53    2    2
 [4,]    0  11    2    3
 [5,]    0  53    2    4
 [6,]    0  53    2    5
 [7,]   22  75    3    4
 [8,]    0  53    3    5
 [9,]   28  81    4    2
[10,]    0  11    4    3
[11,]  279 332    4    7
```

```
> e1[2, 7] ← NA
> e1[3, 7] ← NA
> lp2.hier(e, e1)$lp.out2[, i]
       min max nrow ncol
 [1,]    0  71    1    5
 [2,]  240 311    1    7
 [3,]    0  81    2    2
 [4,]    0  11    2    3
 [5,]    0  75    2    4
 [6,]    0  71    2    5
 [7,]  570 808    2    7
 [8,]    0  75    3    4
 [9,]    0  71    3    5
[10,]   39 185    3    7
[11,]    0  81    4    2
[12,]    0  11    4    3
[13,]  240 332    4    7
```

You can easily compare the protection intervals (on the output of function lp.hier you can see their row number and column number) with the true values. We can see that the table is still not protected.

5.8 Conclusions

Using R for disclosure control has many advantages. The data import/export facilities are very flexibility and powerful and this is really useful in context of disclosure control. With R you can see the perturbation of microdata as a kind of explorative data analysis. We have a powerful system to analyse and evaluate the results and methods during the process of masking data. For all methods there are print, summary and plot methods implemented. Diagnostic plots and plots for the comparison of the original and the perturbed data are very useful during the process of perturbation. Several methods can be evaluated on basis of your data, and diagnostic tools can be used at the same time. The open source packages are highly extendable and can be well documented by use of online-help pages, vignettes and integrated examples. When calculation time is important the code can be written in C or Fortran and can be included in an R package easily. Everybody is invited to contribute to these packages or to make your own R packages for disclosure control.

The robustification of the pca approach for microaggregation with projection pursuit on each cluster leads often to the best results compared with other microaggregation methods.

Also the extension of the single axis method by sorting in each cluster by the most influencial variable can provide good results. Adding noise can be also provide good results and the perturbation must not be applied on all observations but rather on observations which can be identified probalby.

While cell perturbation and controlled tabular adjustment seems to be very good methods for protecting hierarchical tables some statistical agencies will keep hold on traditional suppression methods for a variety of reasons.

In my point of view statistical agencies will have commercial software with guaranteed support for disclosure control or they want to have free available and modifiable open-source software for disclosure control. Having open source code in an attractive software system for doing disclosure control might be a first step forward to harmonize the application of methods for all european statistical agencies.

6 The anonymisation of the CVTS2 and income tax dataset. An approach using R-Package *"sdcMicro"*

To appear in "Monographs of Official Statistics", Eurostat, 2009

Bernhard Meindl*, Matthias Templ*,**

* Department of Methodology, Statistics Austria, Guglgasse 13, 1110 Vienna, Austria. (bernhard.meindl@statistik.gv.at) and

** Department of Statistics and Probability Theory, Vienna University of Technology, Wiedner Hauptstr. 8-10, 1040 Vienna, Austria. (templ@statistik.tuwien.ac.at)

Abstract. The demand for microdata for research and teaching purposes gets higher and higher. To overcome this fact we not only provide secure workstations where researchers can deal with "original" data but also provide more and more anonymised microdata. When using flexible software tools for anonymisation we can provide high quality anonymised data which can be generated in less time. In this paper we show such anonymisation processes on two different data set,

CHAPTER 6. CVTS2 AND INCOME TAX DATA

6.1. INTRODUCTION

the continuing vocational training survey (CVTS2) and the income tax data from 2005.

6.1 Introduction

Since the demand to publish micro data for researchers grows, it is necessary to take actions that published data will not lead to correct re-identification of either an individual or an enterprise. To assure the anonymity of micro data, different anonymisation methods such as global recoding, local suppression or microaggregation are used. Special emphasis should be placed on trying to keep the (multivariate) structure of the data and changing the original data as little as possible while guaranteeing a low individual risk of re-identification. With new and modern software it is possible to use anonymisation methods very effective.

6.1.1 Terms of Use

In Statistics Austria we provide two different variants of anonymised data sets. An SDS (standardised data set) is basically an anonymised micro data set for research and teaching purposes whereas an ADS (task-related dataset) is generated for special research purposes for only one research project or research collaboration with other organisations. However, users have to agree on the terms of use for both variants of anonymised data. From the anonymisation point of view, a SDS is somewhat between a public use file and a scientific use file. The complete terms of use can be found at: http://www.statistik.at/web_de/services/mikrodaten_fuer_forschung_und_lehre/datenangebot/standardisierte_datensaetze_sds/index.html.

6.2 Software used

To protect the CVTS data and the Austrian income tax data we used the R-package sdcMicro (Templ [2007d]). R (R Development Core Team [2008a]) is an open-source, free-available, high-level programming language for statistical computing and graphics. The main advantages of package sdcMicro are the reproducibility of any results, the flexible usage of the package, the import/export facilities, the richness of methods implemented, the graphical power for comparison of original data and perturbed data, the easy usage of the package and the fast calculation of results. Furthermore, one can easily use the whole power of R since the package runs in R.

Since all the results from anonymisation can be reproduced by running a script or parts of the script with the code, a user can do the anonymisation approach very flexible and in an explorative manner and can interactively communicate with all the objects in the workspace of R. It's a bit like playing with the data instead of writing a "batch file" which then must be processed.

6.3 CVTS2 Data

It was the objective to generate a SDS for the CVTS2 dataset (continuing vocational training survey). The goal of this survey is to gain information on internal measures enterprises have taken and (partly) payed for on advanced vocational training for employees. The raw data consisted of 2613 enterprises for which a total of 197 variables has been recorded. It has to be noted that the sample of the CVTS survey is chosen in a way that only enterprises with at least 9 employees may be drawn into the sample. Due to imputation and plausibility checks we observed some abnormalities such as ratios being greater than 100%, most of which can be explained though. Further information on the CVTS2 data can be found

CHAPTER 6. CVTS2 AND INCOME TAX DATA

6.3. CVTS2 DATA

at Statistics Austria's webpage at: http://www.statistik.at/web_de/statistiken/ bildung_und_kultur/erwachsenenbildung_weiterbildung_lebenslanges_lernen/index. html.

The difficulty in generating a SDS for this data was the large number of categorical variables and the fact that any combination of these variables might be used by an attacker to correctly identify an enterprise. Thus it was necessary to assess different scenarios of statistical disclosure control by considering different subsets of the available categorical variables as key variables in order to receive an impression of the disclosure and re-identification risk.

Another important point is that we decided to calculate and publish ratios for most of the numeric variables. Doing so, absolute values can not be recycled anymore. This approach was described as well by Brandt and Hafner (Brandt and Hafner [2007]). However, most multivariate statistical methods are invariant against transformations and the same results can be obtained as for the original values. But, of course, certain aggregations of the data are not suitable for the transformed subset of the data.

6.3.1 Anonymisation of the CVTS2 data

Before actually starting to apply anonymisation methods, 29 variables which were either direct identifiers or were including non relevant information were deleted from the data set. Then ratios for most numeric variables in the dataset were calculated as already explained above. Furthermore, we recoded several variables such as the *economic classification of the enterprise* or the *number of employees* into broader categories. All operations have been done using R and package sdcMicro which make the entire anonymisation procedure flexible, fast and reproducible.

CHAPTER 6. CVTS2 AND INCOME TAX DATA

6.3. CVTS2 DATA

We started by comparing different scenarios and several combinations of possible key variables by having a look at the corresponding individual risks for re-identification Franconi and Polettini [2004b] as well as the number of unique combinations of the characteristics in the key variables. It is in fact very convenient to compare different scenarios using sdcMicro because the user only has to specify the desired key-variables and re-run the code. After comparing several possibilities we decided to use the following key variables which are listed below:

- **economic classification of the enterprise:** 10 categories

- **number of employees:** 4 categories

- **generated revenues for vocational training:** 2 categories

- **expenses for vocational training:** 2 categories

It should be noted that all of these key variables have been already changed in an explorative manner by global recoding techniques or have been generated from other variables.

Then we looked at the number of unique combinations of the key variables, the number of observations with a given combination of the key variables that occurs only twice as well as the individual risk for re-identification. We observed 58 unique observations and 50 observations whose combination of values of the key variables occurred exactly two times.

sdcMicro provides a method to plot the individual risk and to interactively change the threshold value similar to the μ-Argus (Hundepool et al. [2006]) plot method. This is helpful to determine suitable threshold values for the local suppression methods that need to be applied to the key variables. Figure 6.1 shows the individual risk for re-identification along with its empirical distribution function for the original, non perturbated data.

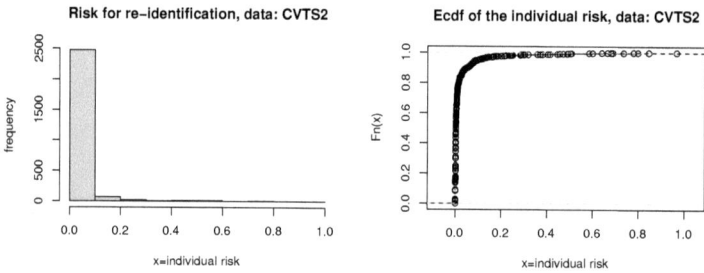

Figure 6.1: individual risk (left) and empirical distribution (right) in original data.

We want to provide k-anonymity (Samarati and Sweeney [1998], P. [2001], Sweeney [2002]) for this dataset. This means that for any combination of key variables at least k observations must exist in the data set sharing that combination. The sdcMicro function localSupp() can be used to suppress values in the key variables. We find 3-anonymity in combination with the other anonymisation methods applied to be a sufficient for publishing for this dataset since the CVTS2 data are not as interesting from an attacker's point of view as for example the income tax data described later.

We started with the variable *beiträge* and set the threshold value to 0.25. This resulted in suppressing 19 values in this key variable. Afterwards, the risk of re-identification of enterprises is plotted again and a new threshold values is determined. Using the threshold value of 0.167, localSupp() was applied to the key variable *einnahmen*. This led to a suppression of 25 values in this variable. The same procedure was used to suppress values in the key variable *a299tot*. After choosing a suitable threshold value (0.104) and applying `localSupp()` we note that 12 values were suppressed in this key variable.

We then observed that there were still enterprises left that had unique combination of the key variables or a combination which only occurred two times in the dataset. Thus, we manually set the values of the variable *beiträge* for those enterprises that had a unique combination of key variables to missing. As a result, 14 suppressions had to be done. Additionally, the variable *einnahmen* was set to missing for all enterprises that had a combination of the key variables that occurred only twice. This resulted in suppressing one additional value. Summarizing this process, a total of 33 values had to be suppressed in variable *beiträge*, 26 variables had to be suppressed in variable *einnahmen* and 12 variables had to be suppressed in variable *a299tot*. After these suppressions, each combination of the values in the key variables occurs at least three times and we note that the goal of 3-anonymity is reached.

In Figure 6.2 the individual risk for re-identification is plotted along with its empirical distribution function after using local suppression for anonymisation. It is obvious, that we achieved a clear reduction in the individual risk of re-identification compared to Figure 6.1 as the different scaling of the *x*-axis indicates.

After dealing with categorical variables and indirect identifiers, we took additional precautions by microaggregating (references on microaggregation can be found in Anwar [1993], Elliot et al. [2005], Defays and Nanopoulos [1993], Defays and M.N. [1998], Domingo-Ferrer and Mateo-Sanz [2002]) the available numeric variables. This effectively means that for each numeric variable the values are grouped by a proximity measure into small groups consisting of 4 values. Then the values within each group are averaged and the aggregate are finally released in the anonymised dataset. This assures that each numeric value occurs at least

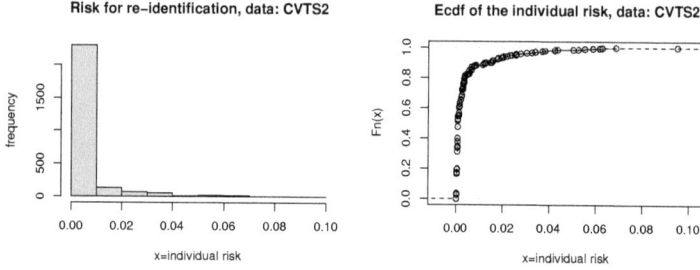

Figure 6.2: individual risk (left) and empirical distribution (right) in anonymised data.

4 times in the anonymised dataset. In this case we have used a version of individual ranking (Defays and Nanopoulos [1993]) which can also be applied on data with missing values.

6.4 Austrian Income Tax Data

The Austrian income dataset for 2005 consists of 5.919.739 rows. The data have already been aggregated from pay slip level to person-level. Thus, exactly one row exists in the raw data for each person that has been liable to pay taxes in 2005. The dataset consists of a total of 74 variables, however, only 17 variables have been included in the published micro data SDS file.

The categorical variables give information about the social status, sex, the federal state and the age of the individual as well as of the number of pay slips considered, the number of weeks employed, whether the person was employed part- or fulltime and the economic classification of the enterprise the individual was predominantly employed at. The quantitative variables included give information for example about the gross wages, social security contributions or information about the amount of other payments a given person has generated.

Further information on the income tax data can be found at the Statistics Austria web page located at:
http://www.statistik.at/web_de/statistiken/oeffentliche_finanzen_und_steuern/_steuerstatistiken/lohnsteuerstatistik/index.html

The final anonymised dataset is available for download at the following web page: http://www.statistik.at/web_de/services/mikrodaten_fuer_forschung_und_lehre/datenangebot/standardisierte_datensaetze_sds/index.html#index8

6.4.1 Anonymisation of the Austrian Income Tax Data

We will now describe the anonymisation methods applied to the raw data set. Our anonymisation approach is quite different than the one used to generate the german public and scientific use file, respectively of income tax data. More on the german contribution in this field can be found on Merz et al. [2005].

First, many quasi direct identifiers and variables that should not be included in the final anonymised data set have been deleted from the raw data. Among the variables deleted are the tax-id which is unique for each person as well as regional information on the individuals such as the exact address. Then, the anonymisation methods described later are applied to the resulting dataset which consisted of all in all 17 variables. Eight variables are scaled categorically while eight variables are quantitative. Furthermore, the sampling weight resulting from drawing a subset of the original data is attached to the dataset as an additional variable.

CHAPTER 6. CVTS2 AND INCOME TAX DATA
6.4. AUSTRIAN INCOME TAX DATA

In a second step, a 1% random sample stratified by age, sex and federal states was drawn from the raw dataset. This resulted in a dataset including 59.279 individuals. This should be seen as a quite effective anonymisation method because even if an attacker manages generate a one to one match from a reference file to an individual from the sample using key variables, he cannot be sure if the possibly identified individual has even been drawn to the sample.

After generating a subset from the raw data it was necessary to define key variables. As already mentioned before, a total of 8 categorical variables was included in the data set. We used sdcMicro to compare different scenarios and several combinations of key variables by having a look at individual risks for re-identification and the number of unique combinations of the characteristics in the key variables. After comparing several possibilities, we considered the following five key variables:

- **social status:** 7 categories
- **federal state:** 10 categories
- **sex:** 2 categories
- **age classes:** 8 categories
- **economic classification of the enterprise:** 12 categories

After deciding on the key variables, the individual risk for re-identification was calculated for all the key variables defined above. It turned out in this explorative approach that we should recode some variables in order to reduce the re-identification risk. Therefore, we recoded variable *social status* as well as the the employment state into less categories. It turned out that 623 observations had a unique combination of characteristics of the key variables and 604 individuals

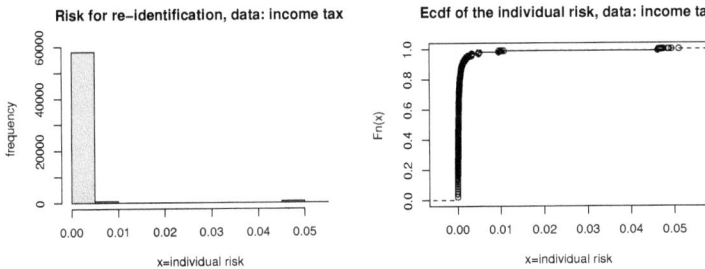

Figure 6.3: individual risk (left) and empirical distribution (right) in original data.

had a combination of values of the key variables that only existed twice. In Figure 6.3 the individual risk of re-identification is plotted for the data using the five key variables discussed above.

In the next step, values of certain key variables are set to missing for individuals with high risk of re-identification. To assess the individual risk and to find a suitable threshold value which determines the number of suppressions we used again the interactive plot method of sdcMicro to assess the individual risk of re-identification and to interactively change the threshold value and look at the resulting re-identification rates conditional on the chosen threshold value.

After choosing a suitable threshold we used the function localSupp() to suppress data in the variable *economic classification of the enterprise* the individuals were employed at with a quite low threshold of 0.01. As a result, a total of 631 values (=≈ 0.01%) for this key variable was set to missing. As a result we obtain that the number of individuals that are unique in the dataset drops to 85 and the number of individuals with a combination of values in the key variables that occurs twice drops down to 256.

After this step, the relative risk is plotted again interactively in order to find a suitable threshold value for local suppression of values in the key variable *social status*. Applying the local suppression method with a threshold value of 0.01 results in no observation that has a unique combination of values in the key variables after setting 93 values in the variable *social status* to missing. However, there are still 154 observation left that have a combination of key values that occurs only twice.

We find that 3-anonymity in addition to microaggregation of numeric variables and the fact that the released data set itself is just a 1% random sample from the population, is adequate for this critical dataset. In order to guarantee 3-anonymity we had to set additional values in the key variables to missing. As already described we decide on a threshold value (0.01) and apply the function `localSupp()` to the key variable *"federal state"*. By setting 154 values of this variable to missing, 3-anonymity for this dataset is obtained.

In Figure 6.4 the individual risk after locally suppressing values in the key variables is plotted. The graph looks similar to Figure 6.3, however one should note the different scaling on the *x*-axis.

After dealing with categorical variables and indirect identifiers, we additionally microaggregated all numeric variables available in the dataset. As for the CVTS2 data we used a version of individual ranking for the microaggregation procedure which guarantees that each numeric value exists at least 4 times in the SDS.

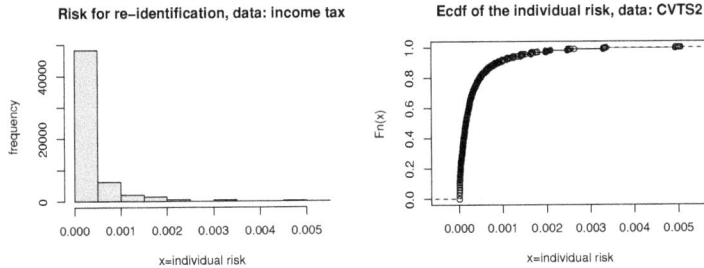

Figure 6.4: individual risk (left) and empirical distribution (right) in anonymised data.

6.5 Conclusion

In both anonymised data sets the re-identification of individuals is very hard or even impossible. For example, each combination of extremely identifiers (Hundepool et al. [2007a]) such as the regional variable occurs more than 1926 times for the income tax data set which itself is only a 1% sample from the original data. Most of the proposed anonymisation rules of the handbook on SDC (Hundepool et al. [2007a]) are not considered since our data are too small to fit this criteria. Nevertheless, global recording, local suppression and microaggregation were applied to achieve sufficient anonymisation of the data, i.e. to provide both 3-anonymity and low re-identification risk. In the other hand, the perturbed data are of high quality and have nearly the same (multivariate) structure as the original data since only few selective recodings and local suppressions on categorical variables were made. It is also well known, that microaggregation does not destroy the (multivatiate) structure of the data (see e.g. in Templ [2006a] or Domingo-Ferrer et al. [2002]). The software used had allowed the anonymisation in less time and in an explorative way.

7 Visualization of Missing Values before Imputation using the R-Package VIM

Matthias Templ[*,**], Andreas Alfons[**], Peter Filzmoser[**]

[*] Department of Methodology, Statistics Austria, Guglgasse 13, 1110 Vienna, Austria. (bernhard.meindl@statistik.gv.at) and

[**] Department of Statistics and Probability Theory, Vienna University of Technology, Wiedner Hauptstr. 8-10, 1040 Vienna, Austria. (templ@statistik.tuwien.ac.at)

Abstract: This paper introduces new tools for the visualization of missing values in R, which can be used for exploring the data and the structure of the missing values. Depending on this structure, they may help to identify the mechanism generating the missing values. This knowledge is necessary for selecting an appropriate imputation method in order to reliably estimate the missing values. Thus the visualization tools should be applied before imputation.

The main goal of our contribution is to stress the importance of visualizing missing values before imputation, and to give a new, easy to use tool into the hand of R users so that visualization of missing values, imputation and data analysis can all be done from within R, without the need of additional software.

All visualization tools presented in this paper are implemented in the R-package **VIM** (**V**isualization and **I**mputation of **M**issing values). A graphical user interface allows an easy handling of the plot methods. **VIM** can be used for data from essentially any field. If spatial coordinates are available, information about missing values can also be displayed in maps.

Using this package, it is possible to explore and analyze the structure of missing values in data and to produce high-quality graphics for publications.

Keywords: Missing Values, Visualization, Missing Values Mechanism, R

7.1 Introduction

Data often contain missing values, and the reasons are manifold [see, e.g., Unwin et al., 1996]. Missing values occur when measurements fail, in case of non-respondents in surveys, when analysis results get lost, or when measurements do not fulfill some prior knowledge of the data, i.e., are implausible (edits). Examples for missing values in the natural sciences are broken measurement units for measurements of ground water quality or temperature, lost soil samples in geochemistry, or soil samples that would need to be re-analyzed but are exhausted. Examples for missing values in official statistics are respondents who deny information about their income, small companies that do not report their turnover, or values that do not fulfill predefined editing rules.

Missing values are often of great interest and must be replaced by meaningful values. Moreover, most of the standard statistical methods can only be applied to complete data, and deleting whole columns or rows of data matrices that contain missing values would result in a loss of important available information. The estimation of missing values is known under the term *imputation* [Little and Rubin, 1987]. In order to be able to choose a proper imputation method, one must be aware of the missing data mechanism(s). The quality of the imputed values

CHAPTER 7. VISUALIZATION OF MISSING VALUES

7.1. INTRODUCTION

depends on the imputation itself and the imputation method used.

Although comprehensive literature on the estimation of missing values is available [e.g., Little and Rubin, 1987, Schafer, 1997], the visualization of missing values is treated only in a few papers [e.g., Unwin, Hawkins, Hofmann, and Siegl, 1996, Eaton, Plaisant, and Drizd, 2005, Cook and Swayne, 2007]. This is also reflected in statistical software. Visualization tools for missing values are rarely or not at all implemented in SAS, SPSS, STATA or even R [R Development Core Team, 2008a]. Through interaction, missing values can be highlighted in GGobi [Cook and Swayne, 2007] and Mondrian [Theus, 2002]. Note that only a few plots are available in GGobi, though. Apple users may be able to use MANET [see, e.g., Unwin, Hawkins, Hofmann, and Siegl, 1996, Theus, Hofmann, Siegl, and Unwin, 1997]. As for GGobi, the power of MANET lies in its interactive features. However, it is restricted to the older *Mac OS* operating system. On earlier versions of the UNIX-based *Mac OS X* operating system for the PowerPC architecture, it is still possible to run MANET in the so-called *Classic Environment* (which is a hardware and software abstraction layer). However, this environment was abandoned in *Mac OS X v10.5*, nor is it available for the newer Intel-based hardware. Another piece of software for missing data analysis is SOLAS, which includes some univariate plots (barcharts, boxplots, normal probability plots, histograms) and a bivariate plot (scatterplot). Unfortunately, it is commercial software and it only runs under Windows XP and Windows Vista operating systems (a demo version is freely available on http://www.statsolusa.com). Table 7.1 lists the available plots for missing values in selected freely available software. For historical reasons, also REGARD [the "forerunner" of MANET, see, e.g., Unwin et al., 1990, Unwin, 1994] is mentioned, as it offers various visualization tools. Furthermore, some visualization tools for missing values in data are implemented in the open-source software ViSta [Young, 1996]. ViSta supports a schematic overview of which values are observed/missing, all existing combinations of observed and

CHAPTER 7. VISUALIZATION OF MISSING VALUES
7.1. INTRODUCTION

missing values in the observations and their frequency, scatterplots with information about missing values in the margins and diamond plots for observations in missing data patterns [see Young et al., 2006]. Only the windows version seems to be up-to-date, and for Apple users, it is restricted to older *Mac OS* operating system.

Since open-source development gets more and more popular, the focus is on open-source software, but also MANET is considered because of its quite powerful visualization techniques for missing values.

It should be pointed out that also the statistical data-visualization system Mondrian has some functionality to plot data with missing values, i.e. it is possible to interactively select missing values via the missing value plot of Mondrian, which is a barchart displaying the proportion of observed and missing values from selected variables. By selecting the missing values within this plot, it is possible to produce plots similar to some of the proposed methods of **VIM**, e.g., histograms, barplots, scatterplots, parallel coordinate plots and scatterplot matrices. However, while Mondrian is a general tool for data visualisation, **VIM** is much more specialised in visualising missing values in the data. With **VIM** it is possible to create high-quality graphics for publications including modifications and additional information added by the user. On the other hand, with Mondrian only screenshots can be taken, and modifying as well as adding to plots is limited. Also in GGobi it is possible to highlight missing values in histograms and parallel coordinate plots through linking.

In this paper, several visualization methods for missing values in R are introduced. Some of them are already implemented in other software environments, some are newly developed. The usefulness of the plots with real data are demonstrated in Section 7.2. Suitable graphical displays of the missing values may also allow the detection of the missing values mechanism (see, e.g., Figure 7.8), which is important for the selection of an appropriate imputation method.

CHAPTER 7. VISUALIZATION OF MISSING VALUES

7.1. INTRODUCTION

All visualization tools presented in this paper are implemented in the R-package **VIM** (**V**isualization and **I**mputation of **M**issing Values), which has been written by the first two authors of this paper. A screenshot of the graphical user interface along with a few comments about the usage of the package is given in Section 7.3.

7.1.1 Missing Value Mechanisms

There are three important cases to distinguish for the responsible generating processes behind missing values [see Rubin, 1976, Little and Rubin, 1987, Schafer, 1997]. The missing values are **M**issing **A**t **R**andom (MAR) if it holds for the probability of missingness that

$$P(X_{miss}|X) = P(X_{miss}|X_{obs}), \qquad (7.1)$$

where $X = (X_{obs}, X_{miss})$ denotes the complete data, and X_{obs} and X_{miss} are the observed and missing parts, respectively. Hence the distribution of missingness does not depend on the missing part X_{miss}.

If the distribution of missingness does not depend on the observed part X_{obs}, the important special case of MAR called **M**issing **C**ompletely **A**t **R**andom (MCAR) is obtained, given by

$$P(X_{miss}|X) = P(X_{miss}).$$

If Equation (7.1) is violated and the patterns of missingness are in some way related to the outcome variables, i.e., the probability of missingness depends on X_{miss}, the missing values are said to be **M**issing **N**ot **A**t **R**andom (MNAR). This relates to the equation

$$P(X_{miss}|X) = P(X_{miss}|(X_{obs}, X_{miss})).$$

Hence the missing values can not be fully explained by the observed part of the data.

CHAPTER 7. VISUALIZATION OF MISSING VALUES

7.1. INTRODUCTION

A practical example for the different missing value mechanisms, which is adequate for the data used in this paper, is given by Little and Rubin [1987]. Considering two variables *age* and *income*, the data are MCAR if the probability of missingness is the same for all individuals, regardless of their age or income. If the probability that income is missing varies according to the age of the respondent, but does not vary according to the income of respondents with the same age, then the missing values in variable income are MAR. On the other hand, if the probability that income is recorded varies according to income for those with the same age, then the missing values in variable income are MNAR [Little and Rubin, 1987]. Naturally, MNAR could hardly be detected (see Section 7.1.2).

Appropriate visualization tools for missing values should be helpful for distinguishing between the three missing value mechanisms. However, there are some limitations that will be described in the following.

7.1.2 Limitations for the Detection of the Missing Values Mechanism

It is often difficult to detect the missing values mechanism in practice exactly, because this would require the knowledge of the missing values themselves [Little and Rubin, 1987]. In the following, a simple example in order to show the limitations for the detection of the missing values mechanism is given.

Figure 7.1 shows a highly correlated bivariate data set. From the complete data (graphic on the left side of the figure) some observations are marked as missing in y (graphic in the middle of the figure) depending on the value of y (the higher y, the higher the probability of missingness). So, the missing values mechanism is constructed as MNAR. In practice, however, only the x-part of the observations with missing values in y is known (graphic on the right of the figure) and can only observe that for increasing x-values the amount of missingness also increases. Therefore, a MAR situation is assumed knowing that this could also be a MNAR situation. In other words, it is impossible to distinguish between

CHAPTER 7. VISUALIZATION OF MISSING VALUES

7.1. INTRODUCTION

MAR and MNAR. From our construction, after taking the relationship between x and the missing data pattern into account, the probability of missingness in y still depends on y (see the graphic in the middle of Figure 7.1, where for approximately equal x-values only high y-values were set to be missing). However, in real world situations a MAR situation is determined for this example, because of the high correlation between x and y [see, e.g., Little and Rubin, 1987]. This results in good estimation by using well-established imputation methods for MAR situations [see, e.g., Dempster et al., 1977].

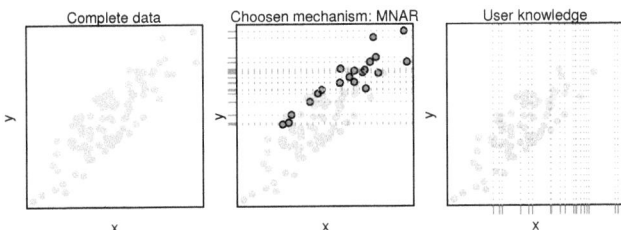

Figure 7.1: Simulated bivariate data set. LEFT: Complete data. MIDDLE: Red points are chosen as missing in y, depending on the value of y (MNAR). RIGHT: Information is only available for x-values in practice.

In Figure 7.2, a similar picture as in Figure 7.1 is obtained, but with uncorrelated variables. Some y-values are again marked as missing, depending on the value of y (MNAR). In practice, however, a MCAR situation is detected, because the missingness seems to be completely independent from the data values. On the other hand, this could also be a MNAR situation, and it is impossible to distinguish between MCAR and MNAR.

To summarize, for a bivariate data set in which one variable contains missing values that are generated with the MNAR mechanism, the following situations

CHAPTER 7. VISUALIZATION OF MISSING VALUES
7.1. INTRODUCTION

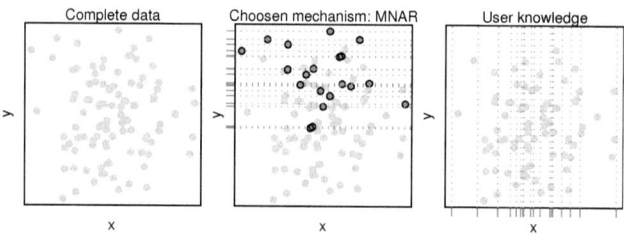

Figure 7.2: Simulated bivariate data set. LEFT: Complete data. MIDDLE: Red points are chosen as missing in y, depending on the value of y (MNAR). RIGHT: Information is only available for x-values in practice.

may occur:

(a) The variables are correlated: MNAR would be classified as MAR, and thus only MCAR and MAR can be detected.

(b) The variables are uncorrelated: MNAR would be classified as MCAR, and thus only MCAR and MAR can be detected.

Multivariate data with missing values in several variables can make it even more complicated to distinguish between the missing value mechanisms. The situation can become even worse in case of outliers, inhomogeneous data or very skewed data distributions.

7.1.3 Detection of the Missing Values Mechanism with Statistical Methods

Consider a bivariate data set with variables (x, y), and let us assume that there are a few missing values in variable y whereas variable x is completely observed. It is of interest whether a MAR situation is present. In principle, a t-test on the x variable where y is the grouping variable (missing or not missing) can be applied, see Little [1988]. If the null hypothesis is rejected under a predefined significance level, a MAR situation can be assumed, because the means of the two groups are significantly different. Such classical tests can lead to meaningless results if the assumptions of the test, i.e., normality of the two samples, independence, and approximately equal variances of the two groups, are not satisfied. However, this is frequently the case when working with real complex data. It can be easily demonstrated that the result of such a test is significant (i.e., indicates MAR) if a clear MCAR situation is constructed and an outlier is included in the data.

Another problem is that the group sizes are in general very different, because the group that represents missing values in y is often much smaller than the group that represents observed data.

To overcome the limitations of the t-test, it is possible to use a robustified generalized linear model in order to predict the missing data mechanism. For a data set with variables $(x_1, \ldots, x_p)$, a binary response variable z can be generated, where 1 stands for missing values in variable x_1 and 0 for the non-missing values in x_1. All other variables of the chosen data set may be the explanatory variables. The error distribution needs to be modeled beforehand as Poisson distributed because of the very different group sizes for variable x_1, i.e., the low probability of missingness.

This approach using robust generalized linear models is quite powerful[1], but adequate (advanced) diagnostic tools have to be used to test whether the model assumptions hold (e.g., multicollinearity or autocorrelation of the residuals). While non-robust generalized linear models again may fail to detect the correct missing values mechanism, the robust version is not affected by outliers. However, it is important to choose an appropriate model, which is often very time consuming, especially when using such complex data like the EU-SILC survey.

---
[1]Robust generalized linear models are already available in R (e.g., in package **robustbase** on `http://cran.r-project.org`)

Therefore, it is strongly recommended to use visualization tools not only to detect the missing value mechanisms, but also to gain insight into the quality and various other aspects of the underlying data.

7.2 Visualization Methods for Missing Values

The visualization tools proposed in this section do not rely on any statistical model assumptions. They are available in the R-package **VIM**, and a simple graphical user interface allows easy handling.

The visualization tools are illustrated on a subset of the European Survey of Income and Living Conditions (EU-SILC) data from 2004 from Statistics Austria (see Table 7.2). This very famous and complex data set is mainly used for measuring poverty and social cohesion in Europe, and for monitoring the Lisbon 2010 strategy of the European Union. The raw data set contains a high amount of missing values, which are imputed with model based imputation methods before public release [Statistics Austria, 2006]. Since a high amount of missing values are not MCAR, the variables to be included for imputation have to be selected carefully. This problem can be solved with our proposed visualization tools.

7.2.1 Aggregation Plot

Often it is of interest how many missing values are contained in each variable. Even more interesting, there may be certain combinations of variables with a high number of missing values. Figure 7.3 displays this information. The barplot on the left hand side shows the number of missing values in each variable. In the *aggregation plot* on the right hand side, all existing combinations of missing and non-missing values in the observations are visualized. A red rectangle indicates missingness in the corresponding variable, a blue rectangle represents available data. Additionally, the frequencies of the different combinations are represented

CHAPTER 7. VISUALIZATION OF MISSING VALUES
7.2. VISUALIZATION METHODS FOR MISSING VALUES

by a small bar plot and by numbers. Variables may be sorted by the number of missing values and combinations by the frequency of occurrence to give more power to finding the structure of missing values. For example, the top row in Figure 7.3 (right) represents the combination with missing values in variables *py010n* (employee cash or near cash income), *py035n* (contributions to individual private pension plans) and *py090n* (unemployment benefits), and observed values in the remaining variables, which appears only once in the data.

The plot reveals an exceptionally high number of missing values in variable *py010n*. The combination with variable *py035n* still contains 32 missing values. Note that it is possible to display proportions of missing values and combinations rather than absolute numbers in Figure 7.3.

7.2.2 Matrix Plot

The *matrix plot* visualizes all cells of the data matrix by rectangles. It is an extension of the function imagmiss() in the R-package **dprep** [Acuna and members of the CASTLE group, 2008]. Red rectangles are drawn for missing values, and a grey scale is used for the available data. To determine the grey level, the variables are first scaled to the interval $[0, 1]$ by subtracting the minimum and dividing by the range. Hence small values are assigned a light grey and high values a dark grey. Additionally, the observations can be sorted by the magnitude of a selected variable, which can also be done interactively by clicking in the corresponding column of the plot.

Figure 7.4 shows a matrix plot of a subset of the EU-SILC data, sorted by variable *pek_n* (net income). It can be seen that the higher the net income, the more missing values exist in variables *py010n* (employee cash or near cash income) and *py050n* (cash benefits or losses from self-employment). Thus the missing data mechanism was found to be MAR for these two variables, which should be considered when applying imputation methods.

CHAPTER 7. VISUALIZATION OF MISSING VALUES
7.2. VISUALIZATION METHODS FOR MISSING VALUES

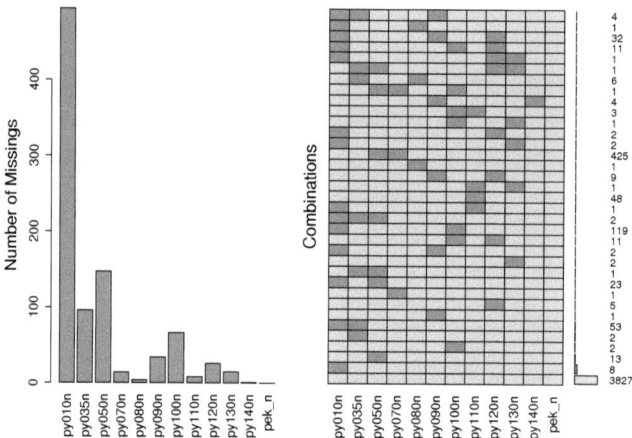

Figure 7.3: Number of missing values for a subsample of the EU-SILC data from Statistics Austria. LEFT: Barplot of the number of missing values in each variable. RIGHT: *Aggregation plot* showing all existing combinations of missing (red) and non-missing (blue) values in the observations, and the corresponding frequencies.

CHAPTER 7. VISUALIZATION OF MISSING VALUES
7.2. VISUALIZATION METHODS FOR MISSING VALUES

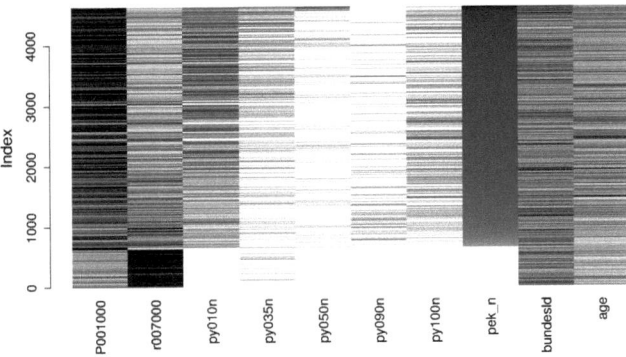

Figure 7.4: Matrix plot of a subset of the EU-SILC data, sorted by variable *pek_n*.

7.2.3 Histogram and Spinogram with Missings

When plotting a histogram of a variable, the amount of missing values in other variables can be shown by splitting each bin into two parts. This is done in Figure 7.5 (left), where the histogram represents the frequencies of the variable *age*, and the bins (age classes) are split according to the numbers of missing (red) and observed (blue) values of variable *py010n* (employee cash or near cash income) for the corresponding age class. Note that missing values can be highlighted for more than one variable (see Section 7.3).

For demonstration purposes, about 15% of the values in *age* were set to be missing. These missing values in *age* are visualized by a small additional barplot on the right side of the plot. The first of these additional bars corresponds to observed values and the second to missing values in *age*. Again, the bars are split according to missing values in *py010n*. As an alternative version, only one bar for missing values in *age* can be drawn instead of the additional barplot. The first option has the advantage that it also provides visual information of missingness in *py010n* for the total amount of observed values in *age*. However, since the additional barplot is on a different scale than the histogram, an extra y-axis is required.

Note that in MANET, a single bar corresponding to the amount of variable *age* would be shown (such as the aforementioned alternative version in **VIM**).

Instead of a histogram, a *spinogram* [Hofmann and Theus, 2005] can also be used. Figure 7.5 (right) shows such a spinogram for the same two variables as used in Figure 7.5 (left). The horizontal axis is scaled according to relative frequencies of the age classes. On the vertical axis, the proportion of missing (bottom) and observed (top) values in *py010n* (employee cash or near cash income) is displayed for each age class. Since the height of each cell corresponds to the proportion of missing/observed values in *py010n*, it is now possible to compare the proportions of missing values among the different age classes. For example, strictly increasing/decreasing proportions for increasing age would indicate a MAR situation. In this spinogram, however, it can only be observed that for age classes > 60 years, the proportion of missing values for employee income becomes very small, which is logical. Analogous to the histogram, a small additional spineplot that visualizes missing values in *age* is drawn. As an alternative version, an additional bar for missing values in *age* can be added to the spinogram instead of the additional spineplot. Similar spinograms are also available in MANET, but without the additional spineplot.

For categorical data, analogous barplots and spineplots with information about missing values are implemented in **VIM**.

7.2.4 Scatterplots

In addition to a standard scatterplot, information about missing values can be displayed. One extension of a scatterplot that visualizes missing values has already been shown in Figure 7.1 (right). A rug representation and red lines are drawn for

CHAPTER 7. VISUALIZATION OF MISSING VALUES
7.2. VISUALIZATION METHODS FOR MISSING VALUES

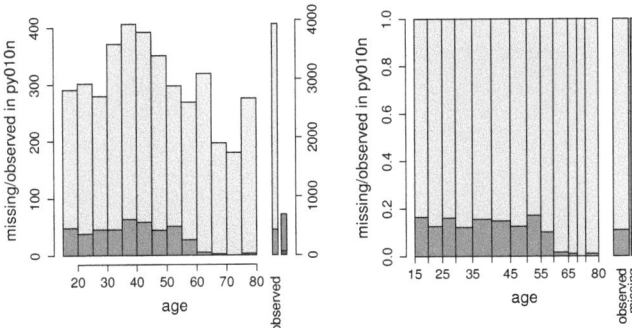

Figure 7.5: Histogram (left) and spinogram (right) of *age* with color coding for missing (red) and available (blue) data in variable *py010n* (employee cash or near cash income).

outliers in one variable. Whether these are drawn for the x- or y-variable can be selected interactively. Optionally, tolerance ellipses can be displayed to indicate the bivariate structure of the data.

Another type of scatterplot is presented in Figure 7.6, using the variables *pek_n* (net income) and *py130n* (disability benefits) of the EU-SILC data. Only observations with values larger than 0 in both variables are used, and both variables are log-transformed (to the base 10). Note that alpha blending is used to prevent overplotting. Moreover, observations with missing values in only one of the variables are shown as univariate scatterplots (red points) along the x- or y-axis, similar to implementations in MANET and GGobi.

However, the implementation in **VIM** also includes boxplots for available (blue) and missing (red) data in the plot margins. Along the horizontal axis, for example, the blue box represents values of the net income where information for disability benefits is also available, whereas the red box represents values of net income where no values for disability benefits are available. Hence, a comparison of the two boxplots can indicate the missing data mechanism. In this example the net income for persons that did not provide information on disability benefits seems to be higher than that for persons who did.

Furthermore, the frequencies of missing values in one or both variables are represented by numbers in the lower left corner.

This kind of bivariate scatterplot can easily be extended to a scatterplot matrix for all combinations of a set of variables. In another version of a scatterplot matrix, observations with missing values in a certain variable or combination of variables are highlighted in the pairwise scatterplots, thus allowing for more than two-dimensional relations. Both versions of scatterplot matrices are available in **VIM**.

7.2.5 Parallel Coordinate Plots with Missings

Parallel coordinate plots show each observation of the scaled data (usually interval scaled to [0,1]) by a line, where the variables are presented by parallel axes [Wegman, 1990]. Similar to previous plots, information about missingness in another variable (or in a combination of variables) is color coded, and alpha blending is used to prevent overplotting. Furthermore, the variables to be used for highlighting may be selected in the main menu of **VIM** or interactively by clicking near the corresponding coordinate axis in the plot. Note that if some variables used for highlighting are also used as plot variables (the variables displayed on the x-axis), disconnected lines may occur. Figure 7.7 shows a parallel coordinate plot of a subset of the EU-SILC data, in which blue lines refer to observed values and red lines to missing values in the variable *py050n* (cash benefits or losses from self-employment).

Clearly, missingness in *py050n* is related with several of the variables used in the plot. Missing values in *py050n* occur primarily in a certain range of the

CHAPTER 7. VISUALIZATION OF MISSING VALUES
7.2. VISUALIZATION METHODS FOR MISSING VALUES

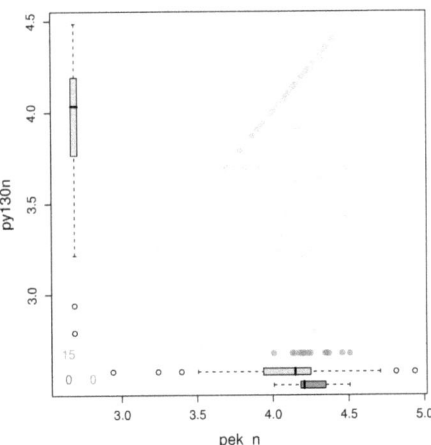

Figure 7.6: Scatterplot of *pek_n* (net income) and *py130n* (disability benefits), both log-transformed, with information about missing values in the plot margins.

variables *pek_g* (gross income) and *age*, and for certain values of *P014000* (profession). Additionally, the amount of missing values depends on the actual values of the variables *P001000* (employment situation) and *bundesld* (region).

A similar implementation of such parallel coordinate plots is available in GGobi.

7.2.6 Parallel Boxplots for Missing Values

Boxplots were already used in Figure 7.6 for comparing the values of one variable for available and missing values in a second variable. This can be extended to a comparison between available and missing data in several variables. Figure 7.8 shows an example for variable *pek_n* (net income), which was log-transformed (to the base 10) after the constant 1 had been added. In addition to a standard boxplot (left), boxplots grouped by observed (blue) and missing (red) values in other variables are drawn. Furthermore, the frequencies of the of the missing values are represented by numbers in the bottom margin. The first line corresponds to the observed values in *pek_n* and their distribution among the different groups, the second line to the missing values.

For some variables, the presence of missing values clearly depends on the magnitude of the values of *pek_n*. For example, missing values in variable *py080n* (pension from individual private plans) occur particularly for high values of the net income. This indicates a MAR situation for missing values in *py080n*.

These boxplots provide a good overview about possible MCAR/MAR situations of one variable according to the other $p-1$ variables. After obtaining this information, one may be able to find a suitable model for imputation of the missing values in the considered variable.

7.2.7 Plot Missings in Maps

If geographical coordinates are available for a data set, it can be of interest to check whether missingness in a variable corresponds to spatial patterns in a map.

CHAPTER 7. VISUALIZATION OF MISSING VALUES
7.2. VISUALIZATION METHODS FOR MISSING VALUES

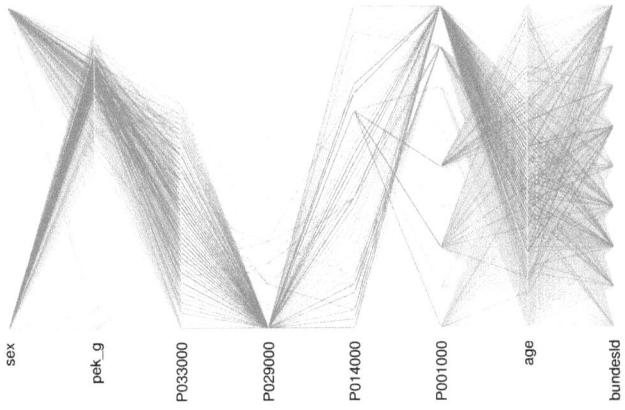

Figure 7.7: Parallel coordinate plot for a subset of the EU-SILC data. Red lines indicate missing values in variable *py050n* (cash benefits or losses from self-employment).

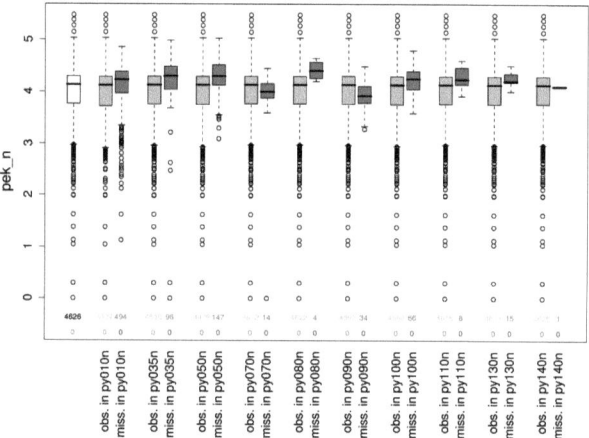

Figure 7.8: The transformed values of variable *pek_n* (net income) are grouped according to missingness in several variables of the EU-SILC data and presented in parallel boxplots.

For example, the values of one variable can be drawn by growing dots in the map, reflecting the magnitude of the data values. Missingness in a second variable (or a combination of variables) can be color coded, and alpha blending can be used to prevent overplotting. This allows conclusions about the relation of missingness to both the values of the variable of interest and the spatial location.

In the example in Figure 7.9, the variable *Ca* of the Kola C-horizon data [Reimann et al., 2008] is displayed with information about missing values in the chemical elements *Bi* and *As*. It shows that the missing values have a regional dependency, i.e., they occur mainly in a certain part of the Kola project area.

The observations in the EU-SILC data are only assigned to one of the nine political regions of Austria, they do not have spatial coordinates. However, the proportion of missing values or the absolute amount of missing values in a variable can still be visualized for the political regions. In Figure 7.10, the proportions of missing values in variable *py050n* (cash benefits or losses from self-employment) are coded according to an *rgb* color scheme, resulting in a higher portion of red for regions with a higher proportion of missing values. Additionally, the proportions of missing values are shown in the regions.

7.3 R-Package VIM

All tools presented in Section 7.2 for visualizing missing values were implemented in the R-package **VIM**. The figures in this paper were produced with **VIM** version 1.3 and R version 2.8.1. Since the development of this software is an ongoing work, it is highly recommended to always use the latest available version from the comprehensive R archive network (CRAN, http://cran.r-project.org).

A graphical user interface (GUI), which was developed using the R-package **tcltk** [R Development Core Team, 2008a], allows easy handling of the functions. Figure 7.11 shows the **VIM** GUI.

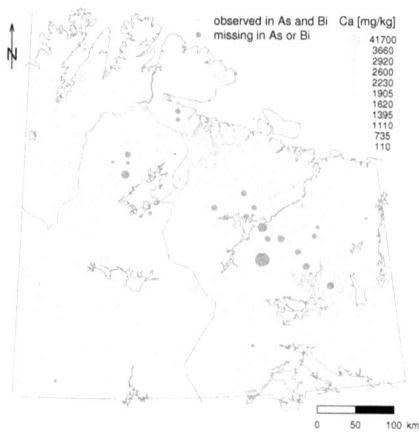

Figure 7.9: Map of the Kola region. Missings in As or Bi show a spatial dependency.

CHAPTER 7. VISUALIZATION OF MISSING VALUES

7.3. R-PACKAGE VIM

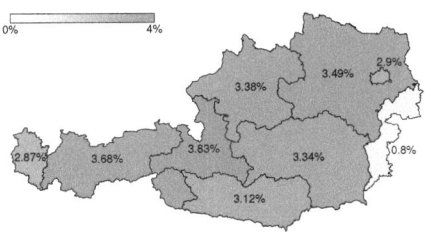

Figure 7.10: Map of the nine political regions of Austria. Regions with a higher proportion of missing values (see the percentages in each region) in variable *py050n* (cash benefits or losses from self-employment) receive a higher portion of red. Further information is provided when clicking on a region in the map.

For visualization, only the *Data*, *Visualization* and *Options* menus are important. The *Data* menu allows to select a data set from the R workspace or load data into the workspace from RData files. Furthermore, it can be used to transform variables, which are then appended to the data set in use. Commonly used transformations in official statistics are available, e.g., the Box-Cox transformation [Box and Cox, 1964] and the log-transformation as an important special case of the Box-Cox transformation. In addition, several other transformations that are frequently used for compositional data [Aitchison, 1986] are implemented. Background maps and coordinates for spatial data can be selected in the data menu as well.

After a data set was chosen, variables can be selected in the main menu, along with a method for scaling. An important feature is that the variables will be used in the same order as they were selected, which is especially useful for parallel coordinate plots. Variables for highlighting are distinguished from the plot variables and can be selected separately. For more than one variable chosen for highlighting, it is possible to select whether observations with missing values in any or in all of these variables should be highlighted.

A plot method can be selected from the *Visualization* menu. Note that plots that are not applicable to the selected variables are disabled, for example, if only one plot variable is selected, multivariate plots cannot be chosen.

Last, but not least, the *Options* menu allows to set the colors and alpha channel to be used in the plots. In addition, it contains an option to embed multivariate plots in Tcl/Tk windows. This is useful if the number of observations and/or variables is large, because scrollbars allow to move from one part of the plot to another.

Interactive features are implemented in the plot methods. There are, however, limited possibilities for interactive graphics in standard R. When variables are selected for highlighting in univariate plots such as histograms, barplots, spineplots or spinograms, it is possible to switch between the variables, i.e., for histograms, clicking in the right plot margin corresponds with creating a histogram for the next variable, and clicking in the left margin switches to the previous variable. This interactive feature is particularly usedful for parallel boxplots, as it allows to view all possible $p(p-1)$ combinations with $p-1$ clicks, where p denotes the number of variables.

For multivariate plots (scatterplot matrix and parallel coordinate plot), variables for highlighting can be selected and deselected interactively, by clicking in a diagonal panel in the scatterplot matrix or on a coordinate axis in the parallel coordinate plot. Information about the current selection is printed on the R-console.

The matrixplot is particulary powerful if the observations are sorted by a specific variable (see Figure 7.4). This can be done by clicking on the corresponding column.

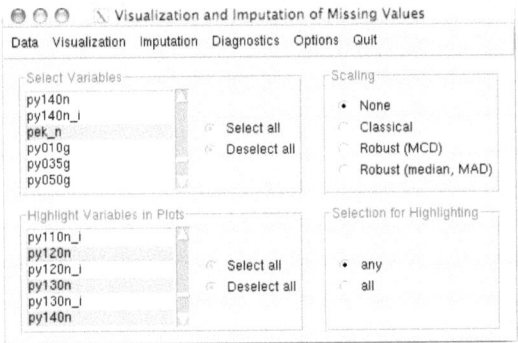

Figure 7.11: The **VIM** GUI.

7.4 Conclusions

Detecting missing value mechanisms is usually done by statistical tests or models. Visualization of missing values can support the test decision, but also reveals more details about the data structure. Most notably, statistical requirements for a test can be checked graphically, and problems like outliers or skewed data distributions can be discovered.

A graphical user interface that provides several possibilities for visualizing data with missing values was developed and implemented in R. The implemented plots allow to combine information about the data with information about missingness in a certain variable or a certain combination of variables. Some plots also allow for an interactive handling. The information resulting from the different graphics can be used for detecting missing value mechanisms, and thus for selecting appropriate methods for data imputation. **VIM** can easily be used within R without the need to install additional software. Moreover, users have the possibility to use the whole power of a statistical system like R at the same time. Therefore, even the re-implementation of some plot methods might be of high interest for the users. In addition, several new plots give a deep insight into the structure of the data and the structure of missing values.

VIM can be used for data from essentially any field. If spatial coordinates are available, it is possible to load a background map and present information about missing values on that map. Spatial patterns of missingness can be very instructive for example in environmental science.

Using **VIM**, it is thus possible to explore and analyze the structure of missing values in data, as well as to produce high-quality graphics for publications.

The package **VIM** is available on the comprehensive R archive network (CRAN, http://cran.r-project.org).

7.5 Acknowledgments

We would like to thank two anonymous referees for their constructive reports.

This work was partly funded by the European Union (represented by the European Commission) within the 7th framework programme for research (Theme 8, Socio-Economic Sciences and Humanities, Project AMELI (Advanced Methodology for European Laeken Indicators), Grant Agreement No. 217322).

Table 7.1: Implementations of plots related to missing values in selected software.

	R	GGobi	VIM	MANET
Platform	all	all	all	Mac OS
Open-source	✔	✔	✔	
Frequency plot			✔	✔
Barplot	✔		✔	✔
Histogram		✔[a]	✔	✔
Spineplot			✔	✔
Spinogram			✔	✔
Mosaicplot			[b]	✔
Boxplot			✔	✔
Scatterplot		✔	✔[c]	✔
Scatterplot matrix		✔	✔[c]	
Matrixplot	✔		✔[d]	
Parallel coordinate plot		✔[a]	✔	
Parallel boxplots			✔	✔
Maps with polygons			✔	✔
Maps with points			✔	
GUI		✔	✔	✔

[a] ... through linking.

[b] ... available soon.

[c] ... more than one version is available in VIM.

[d] ... a more sophisticated version is implemented in VIM.

Table 7.2: Explanation of the used variables from the EU-SILC data set.

name	meaning
py010n	employee cash or near cash income
py035n	contributions to individual private pension plans
py050n	cash benefits or losses from self-employment
py070n	values of goods produced by own-consumption
py080n	pension from individual private plans
py090n	unemployment benefits
py100n	old-age benefits
py110n	survivors' benefits
py120n	sickness benefits
py130n	disability benefits
py140n	education-related allowances
pek_n	net income
pek_g	gross income
P001000	employment situation
r007000	occupation
P033000	years of employment
P029000	hours worked per week, miscellaneous employment
P014000	profession
bundesld	region
age	age
sex	sex

8 Imputation of missing values for compositional data using classical and robust methods

Re-submitted to the "Journal of Computational Statistics and Data Analysis" in April 2009.

Karel Hron*, Matthias Templ**,***, Peter Filzmoser***

* Department of Mathematical Analysis and Applications of Mathematics, Palacký University, Tomkova 40, 779 00 Olomouc, Czech Republic

** Department of Methodology, Statistics Austria, Guglgasse 13, 1110 Vienna, Austria. (matthias.templ@statistik.gv.at) and

*** Department of Statistics and Probability Theory, Vienna University of Technology, Wiedner Hauptstr. 8-10, 1040 Vienna, Austria. (templ@statistik.tuwien.ac.at)

Abstract: Compositional data are multivariate data consisting of the parts of some whole, conveying exclusively relative information. Examples are the concentration of chemical elements of soil samples, or the household expenditures in different commodity groups. The fact that compositional data contain only relative information has to be taken into account also for imputation methods.

CHAPTER 8. IMPUTATION OF MISSING VALUES FOR CODA
8.1. INTRODUCTION

New imputation algorithms for estimating missing values in compositional data are introduced. A first proposal uses the k-nearest neighbor procedure based on the Aitchison distance, a distance measure especially designed for compositional data. It is important to adjust the estimated missing values to the overall size of the compositional parts of the neighbors. As a second proposal an iterative model-based imputation technique is introduced which initially starts from the result of the proposed k-nearest neighbor procedure. The method is based on iterative regressions, hereby accounting for the whole multivariate data information. The regressions have to be performed in a transformed space, and depending on the data quality classical or robust regression techniques can be employed. The proposed methods are tested on a real and on simulated data sets. The results show that the proposed methods outperform standard imputation methods. In presence of outliers the model-based method with robust regressions is preferable.

Keywords: missing values, logratio transformations, balances, robust regression, k-nearest neighbor methods

8.1 Introduction

Most statistical methods cannot directly be applied to data sets including missing observations. While in the univariate case the observations with missing information could simply be deleted, this can result in a severe loss of information in the multivariate case. Multivariate observations usually form the rows of a data matrix, and deleting an entire row implies that cells carrying available information are lost for the analysis. In both cases (univariate and multivariate), the problem remains that valid inferences can only be made if the missing data are *missing completely at random* (MCAR) [see, e.g., Little and Rubin, 1987]. Instead of deleting observations with missing values it is thus better to fill in the missing cells with appropriate values. This is only possible if additional information is available, i.e.

only in the multivariate case. Once all missing values have been imputed, the data set can be analyzed using the standard techniques for complete data.

Many different methods for imputation have been developed over the last few decades. While univariate methods replace the missing values by the coordinate-wise mean or median, the more advisable multivariate methods are based on similarities among the objects and/or variables. A typical distance based method is k-nearest neighbor ($k$nn) imputation, where the information of the nearest $k \geq 1$ complete observations is used to estimate the missing values. Another well-known procedure is the EM (expectation maximization) algorithm [Dempster et al., 1977], which uses the relations between observations and variables for estimating the missing cells in a data matrix. Further details, as well as methods based on multiple regression and principal component analysis are described in Little and Rubin [1987] and Schafer [1997]. Most of these methods can deal with both, MCAR and *missing at random* (MAR) missing values mechanisms [see, e.g., Little and Rubin, 1987]. Moreover, one usually assumes that the data originate from a multivariate normal distribution, which is no longer valid in presence of outliers in the data. In this case the "classical" methods can give very biased estimates for the missing values, and it is more advisable to use robust methods, being less influenced by outlying observations [see, e.g., Beguin and Hulliger, 2008, Serneels and Verdonck, 2008]. Classical or robust imputation methods turned out to work well for standard multivariate data, i.e. for data with a direct representation in the Euclidean space. This, however, is not the case for compositional data, and thus a different approach for imputation has to be used.

Compositional data occur frequently in official statistics (tax components in tax data, income components, wage components, expenditures, etc.), in environmental and technical sciences, and in many other fields. An observation $x = (x_1, \ldots, x_D)^t$ is by definition a D-part composition if, and only if, all its components are strictly positive real numbers, and if all the relevant information is contained in the ratios between them [Aitchison, 1986]. As a consequence of this formal definition, $(x_1, \ldots, x_D)^t$ and its $c > 0$ multiple $(cx_1, \ldots, cx_D)^t$ contain essentially the same information. A typical example for compositional data are data arising from a chemical analysis of a sample material. The essential information is contained in the relative amounts of the element concentrations, and not in the absolute amounts which would depend on the weight of the sample material. One can thus define the *simplex*, which is the sample space of D-part compositions, as

$$\boldsymbol{x} = (x_1, \ldots, x_D)^t, \quad x_i > 0, \quad i = 1, \ldots, D, \quad \sum_{i=1}^{D} x_i = \kappa. \tag{8.1}$$

The constant κ represents the sum of the parts. Since only ratios between the parts are of interest, κ can be chosen as 1 or 100, because then the parts of a composition can be interpreted as probabilities or percentages. Note that the constant sum constraint implies that D-part compositions are only $D-1$ dimensional, so they are singular by definition. This causes limitations for the statistical analysis, but the fact that compositional data have no direct representation in the Euclidean space but only in the simplex sample space has even more severe consequences. Applying standard statistical methods like correlation analysis or principal component analysis directly to compositional data would give misleading results [Pearson, 1897, Filzmoser and Hron, 2008b, Filzmoser et al., 2008]. This is also true for imputation methods [Bren et al., 2008, Martín-Fernández et al., 2003, Boogaart et al., 2006].

In this paper we introduce procedures to estimate missing values in compositional data. For this purpose, more details on the nature and geometry of compositional data have to be provided in Section 2. Section 3 introduces an algorithm based on the k-nearest neighbor technique, and an iterative algorithm for the estimation of missings in compositional data. A modification of the latter algorithm will allow to deal with data that are contaminated by outliers. A small data example in Section 4 demonstrates the usefulness of the new routines. In simulation studies in Section 5 the new procedures are compared with standard imputation methods that are directly applied to the raw compositional data. The final Section 6 concludes.

8.2 Further properties of compositional data

Although compositional data are characterized by the constant sum constraint, the value of κ in Equation (8.1) can be different for different observations. For example, when sample materials are only analyzed for some chemical elements but not analyzed completely, the sum of the element concentrations of the different samples will in general not be the same. This, however, should not affect the imputation method, because all the relevant information is contained in the ratios between the parts of the observations.

An example is shown in Figure 8.1 (left). Each data point consists of two compositional parts. The dashed line indicates the constant sum constraint, and according to Equation (8.1) it is chosen at $\kappa = 1$. The sum of the compositional parts of the data points is smaller than κ. Each point could be shifted along the line from the origin through the point without changing the ratio of the two compositional parts. More formally, an observed composition $x = (x_1, \ldots, x_D)^t$ is defined as a member of the corresponding *equivalence class* of x,

$$\underline{x} = \{cx, \, c \in \mathbf{R}^+\}.$$

Thus two compositions which are elements of the same equivalence class $\underline{x}$, contain the same information and they are also called compositionally equivalent [Pawlowsky-Glahn et al., 2005]. Therefore we could even project the data points along the lines from the origin to the dashed line without changing the information of the compositional data. This fact has to be considered for an appropriate choice of a distance measure.

CHAPTER 8. IMPUTATION OF MISSING VALUES FOR CODA
8.2. FURTHER PROPERTIES OF COMPOSITIONAL DATA

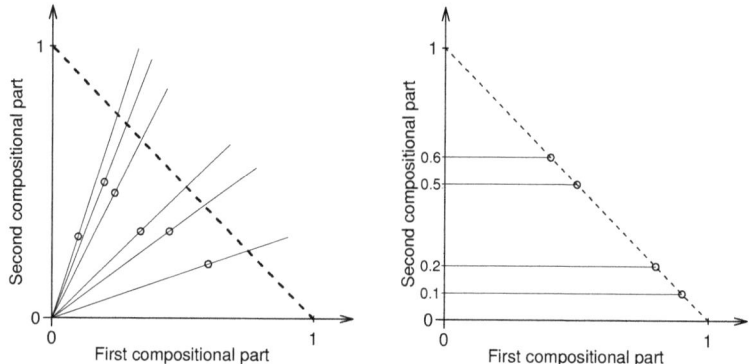

Figure 8.1: Left plot: Two-part compositional data without the constraint of constant sum. The points could be varied along the lines from the origin without changing the ratio of the compositional parts. Right plot: According to the relative scale, the points close to the boundary are more different than the central points, although the Euclidean distances are the same. The Aitchison distance accounts for this fact.

A further geometrical peculiarity of compositional data is that data points close to the boundary of the sample space are related in a different way than data points in the center. This fact has to be considered for example in outlier detection [Filzmoser and Hron, 2008a], but also for designing a distance measure. Figure 8.1 (right) shows for two-part compositional data two data points close to the boundary and two points in the center of the sample space. Although the data pairs visually have the same distance (because we are thinking in terms of Euclidean distances), the increase along the vertical axis of the boundary points is much larger than that of the points in the center (for the boundary points the increase is by a factor 2 from 0.1 to 0.2, whereas for the central points the increase is only by a factor 1.2 from 0.5 to 0.6). A distance measure that is accounting for this relative scale property is the Aitchison distance [Aitchison et al., 2000], defined for two compositions $x = (x_1, \ldots, x_D)^t$ and $y = (y_1, \ldots, y_D)^t$ as

$$d_A(x, y) = \sqrt{\frac{1}{D} \sum_{i=1}^{D-1} \sum_{j=i+1}^{D} \left(\ln \frac{x_i}{x_j} - \ln \frac{y_i}{y_j} \right)^2}. \tag{8.2}$$

As an example, the two boundary points in Figure 8.1 (right) have an Aitchison distance of 0.57, whereas the two central points have Aitchison distance 0.29.

Replacing the Euclidean distance by the Aitchison distance is necessary because the simplex sample space has a different geometrical structure than the classical Euclidean space. Principles on this geometry were introduced in Aitchison [1986], and the resulting so-called Aitchison geometry holds the vector space as well as Hilbert space properties [see, e.g., Egozcue and Pawlowsky-Glahn, 2006]. This allows to construct a basis on the simplex, and consequently standard statistical methods designed for the Euclidean space can be used. Out of several proposals for the construction of a basis [Pawlowsky-Glahn et al., 2005], the isometric logratio (ilr) transformation [Egozcue et al., 2003] seems to be the most convenient one. The ilr transformation results in a $D - 1$ dimensional real space, and it offers good theoretical and practical properties [Egozcue and Pawlowsky-Glahn, 2005]. One important property is the *isometry*, meaning that the Aitchison distance of two compositions x and y is the same as the ordinary Euclidean distance d_E for their ilr images $ilr(x)$ and $ilr(y)$, i.e.

$$d_A(x, y) = d_E(ilr(x), ilr(y)). \tag{8.3}$$

Thus, the ilr transformation allows to represent compositional data in terms of the standard Euclidean geometry, and therefore standard statistical methods can be applied. Note that this property is also fulfilled for the centered logratio (clr) transformation [Aitchison, 1986], but this transformation results in singular data, causing problems for robust estimation. The third well-known logratio transformation, the additive logratio (alr) transformation Aitchison [1986] also yields a $(D-1)$-dimensional real space. This transformation, however, is not isometric and thus not recommended for analyzing compositional data [e.g. Pawlowsky-Glahn et al., 2005].

An example is shown in Figure 8.2. The original two-part compositional data are shown in the left plot with symbols ∘, △, and +. Introducing a constant sum constraint would correspond to projecting the data as indicated, resulting in data with the filled symbols. The ilr transformation reduces the dimensionality by one, and the resulting univariate data are shown on the right-hand side of Figure 8.2. As it can be seen, the ilr transformation for the original data (upper part) is the same as for the data scaled to constant sum (lower part). The ilr transformed data clearly reveal an outlier group, originating from the data points + in the left plot. Although Figure 8.2 (left) shows three data clouds, only the cloud with symbols + forms outliers, because the other two clouds cannot be distinguished when thinking in terms of equivalence classes. This fact has to be considered also for the estimation of missing parts in compositional data.

The ilr transformation raises a problem because the new coordinates (often

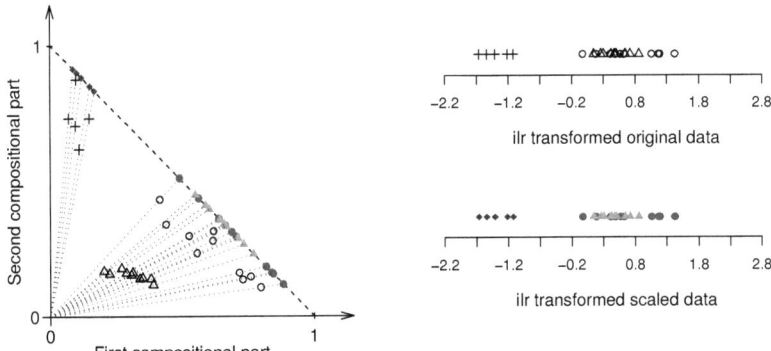

Figure 8.2: Left plot: Two-part compositional data consisting of three groups, and their projection on the line indicating constant sum 1. While the relative information of the groups with symbols ○ and △ is similar, the data points of the group with symbol + contain very different information. Right plot: In the upper part the ilr transformed original data are shown. The lower plot shows the ilr transformed data with constant sum constraint. This demonstrates that the constant sum constraint does not change the ilr transformed data.

called *balances*) have no interpretation in the sense of the original compositional parts. This is due to the definition of compositional data which contain only relative information, thus making a meaningful interpretation of such variables impossible. A possible solution is to split the parts into separated groups and to construct balances representing the groups and balances representing the relations between the groups. This construction procedure is called *sequential binary partitioning* [Egozcue and Pawlowsky-Glahn, 2005]. The resulting balances are in fact just orthogonal rotations, but they have the advantage that groups of variables can be directly assigned to groups of balances.

Sequential binary partitioning is also useful in the context of estimating missing values. For example, if the missing values are mainly contained in the first compositional part of the data, one can choose the ilr transformation as

$$ilr(\boldsymbol{x}) = (z_1, \ldots, z_{D-1})^t, \quad z_j = \sqrt{\frac{D-j}{D-j+1}} \ln \frac{\sqrt[D-j]{\prod_{l=j+1}^{D} x_l}}{x_j}, \quad \text{for } j = 1, \ldots, D-1 .$$
(8.4)

This choice of the balances separates all the relative information of the part x_1 from the remaining parts $x_2, \ldots, x_D$. The balance z_1 contains all the relative information of part x_1 to all the remaining parts, represented by $z_2, \ldots, z_{D-1}$. This choice of the balances will be very useful for estimating missing values in x_1 by regression on the remaining variables, see Section 3.

8.3 Imputation methods for compositional data

In Martín-Fernández et al. [2003] the estimation of missing values in compositional data was done in the sense of the Aitchison geometry, but with the constraint of constant sum of the parts. We follow directly the definition of compositions by considering only ratios between the parts. This is realistic because compositional data include the information in the ratios, and in many practical cases the sum of the parts is not constant (see the example in Section 4). If the constant sum constraint is required, one can divide all data values (the observed and the imputed values) by their sum and multiply by the desired constant. Alternatively, Martín-Fernández et al. [2003] suggested to modify only the nonmissing values.

The easiest way to impute missing values in compositional data is to replace the missing value of a part by the geometric mean of all available data in this part. Here, one has to correct the geometric mean by the ratio of the sum of the parts of the incomplete observation and the sum of the elements of the geometric mean vector [center of compositional data set, see Pawlowsky-Glahn and Egozcue, 2002]. This approach, however, does not account for the multivariate data structure, although it is often used for imputation of missing values for compositional data sets. Another approach was introduced by Palarea-Albaladejo and Martín-Fernández [2008]. Here the EM-algorithm [Dempster et al., 1977] for alr-transformed compositional data was used for the imputation. Although this algorithm was originally introduced for replacing rounded zeros, it can be adapted for estimating missing values. This approach, however, is not robust against outlying observations, and the alr-transformation is not isometric. Thus, robustifying this method may not improve the imputation because of the lack of good geometrical properties. In the following, this algorithm will be denoted by alr-EM.

In the following we present two new approaches that use the multivariate data information for imputation.

8.3.1 k-nearest neighbor ($k$nn) imputation

$k$nn imputation turned out to be successful for standard multivariate data [Troyanskaya et al., 2001]. The idea is to use a distance measure for finding the k most similar observations to a composition containing missings, and to replace the missings by using the available variable information of the neighbors. In the

CHAPTER 8. IMPUTATION OF MISSING VALUES FOR CODA
8.3. IMPUTATION METHODS FOR COMPOSITIONAL DATA

context of compositional data we have to use an appropriate distance measure, like the Aitchison distance (Section 2).

Suppose that a composition contains missing values in several cells. Then the imputation can be done

(1) simultaneously for all cells, by

   a) searching the k-nearest neighbors among all complete observations,

   b) searching the k-nearest neighbors among observations which may be incomplete, but where the information in the variables to be imputed plus some additional information is available;

(2) sequentially (one cell after the other), by

   a) searching the k-nearest neighbors among observations where all information corresponding to the non-missing cells plus the information in the variable to be imputed is available,

   b) searching the k-nearest neighbors among observations where in addition to the variable to be imputed any further cells are non-missing.

For estimating the missing parts of a composition, the approaches in (1) use the information of the same k observations, whereas for the approaches in (2) the k observations can change during the sequential imputation. In the following we will use the approach (2a) because in general more neighbors will be considered for imputation, and requesting more information per observation will lead to a more reliable imputation result.

For imputing a missing part of a composition we use the median of the corresponding cells of the k-nearest neighbors. However, we first have to adjust the cells according to the overall size of the parts. This was not necessary for finding the k-nearest neighbors, because the Aitchison distance is the same for any compositions x and y belonging to equivalence classes $\underline{x}$ and $\underline{y}$ (see Section 2).

More formally, let us consider a composition $x_i = (x_{i1}, \ldots, x_{iD})^t$, $i = 1, \ldots, n$, with n the number of observations, and let $M_i \subset \{1, \ldots, D\}$ denote the set of indexes referring to the missing cells of x_i. Then $O_i = \{1, \ldots, D\} \backslash M_i$ refers to the observed parts of x_i. For imputing a missing cell x_{ij}, for any $j \in M_i$, we consider among all remaining compositions those which have non-missing parts at positions j and O_i, and compute the k-nearest neighbors $x_{i_1}, \ldots, x_{i_k}$ to the composition x_i using the Aitchison distance. The j-th cell of all k-nearest neighbors is of interest for imputation. First we have to adjust these cells by factors comparing the size of the parts in O_i. The adjustment factors can be taken as

$$f_{ii_l} = \frac{\sum_{o \in O_i} x_{io}}{\sum_{o \in O_i} x_{i_l o}} \quad \text{for } l = 1, \ldots, k. \tag{8.5}$$

Using these factors as weights for the observations will make the k-nearest neighbors comparable. The imputed value replacing the missing cell x_{ij} is

$$x_{ij}^* = \text{median}\{f_{ii_1} x_{i_1 j}, \ldots, f_{ii_k} x_{i_k j}\}. \tag{8.6}$$

By taking the median we obtain robustness to outliers in the j-th parts of the k-nearest neighbors.

Although the choice of the adjustments in (8.5) is coherent with the definition (8.1) of compositional data, a more robust version could be preferable. In the example and in the simulations below we will thus use the adjustment factors

$$f_{ii_l}^* = \frac{\underset{o \in O_i}{\text{median}}\, x_{io}}{\underset{o \in O_i}{\text{median}}\, x_{i_l o}} \quad \text{for } l = 1, \ldots, k, \tag{8.7}$$

which will lead to more stable results for contaminated data.

$k$nn imputation is numerically stable (no iterative scheme is required), but it has some limitations. First of all, the optimal number k of nearest neighbors has to be determined. Ideally, this number can be found within a simulation, by randomly setting observed cells to missing, estimating these missings based on different choices for the number k, and measuring the error between the imputed and the originally observed values. The k producing the smallest error can be considered as optimal. Secondly, $k$nn imputation does not fully account for the multivariate relations between the compositional parts. This is only considered indirectly when searching for the k-nearest neighbors.

From this point of view, the quality of the imputation may be improved by a model-based imputation procedure, as introduced in the following section.

8.3.2 Iterative model-based imputation

Our focus is on an iterative regression-based procedure. In each step of the iteration, one variable is used as a response variable and the remaining variables serve as the regressors. Thus the multivariate information will be used for imputation in the response variable. Since we deal with compositional data we cannot directly use the original data in regression, but we have to work in a transformed space. For this purpose we choose the ilr transformation with the concept based on balances, because of its advantageous properties. However, already for constructing the balances a data matrix with complete information is needed, see Equation (8.4). This can be overcome by initializing the missing values with $k$nn imputation, as described above. A further difficulty is that several (or even all) variables have to be used for constructing a balance. Thus, if the initialization of the missings was poor, one can expect a kind of error propagation effect. In order to avoid this, we have to choose the balances carefully. The choice of the balances by Equation (8.4) is an attempt in achieving the highest possible stability with respect to missing values. For example, the missing values that are replaced in the first variable x_1 will only affect the first balance z_1, but they have no influence on the remaining balances. Thus, using such a sequential binary partition will cause that as few as possible balances are affected by the missing values.

Considering a data matrix with n observations and D parts, Equation (8.4) can be rewritten for the i-th composition $\boldsymbol{x}_i = (x_{i1}, \ldots, x_{iD})^t$, $i = 1, \ldots, n$, as $ilr(\boldsymbol{x}_i) = \boldsymbol{z}_i = (z_{i1}, \ldots, z_{i,D-1})^t$, where

$$z_{ij} = \sqrt{\frac{D-j}{D-j+1}} \ln \frac{\sqrt[D-j]{\prod_{l=j+1}^{D} x_{il}}}{x_{ij}}, \quad \text{for } j = 1, \ldots, D-1 \ . \quad (8.8)$$

The corresponding inverse transformation is $ilr^{-1}(\boldsymbol{z}_i) = \boldsymbol{x}_i = (x_{i1}, \ldots, x_{iD})^t$, with

$$x_{i1} = \exp\left(-\frac{\sqrt{D-1}}{\sqrt{D}} z_{i1}\right), \quad (8.9)$$

$$x_{ij} = \exp\left(\sum_{l=1}^{j-1} \frac{1}{\sqrt{(D-l+1)(D-l)}} z_{il} - \frac{\sqrt{D-j}}{\sqrt{D-j+1}} z_{ij}\right), \quad \text{for } j = 2, \ldots, D \quad (8.10)$$

$$x_{iD} = \exp\left(\sum_{l=1}^{D-1} \frac{1}{\sqrt{(D-l+1)(D-l)}} z_{il}\right) \quad (8.11)$$

(with possible normalization of the compositional parts to a chosen constant sum).

An iterative algorithm based on regression can be summarized as follows:

Step 1: Initialize the missing values using the $k$nn algorithm based on Aitchison distances, as described above.

Step 2: Sort the variables according to the amount of missing values. In order to avoid complicated notation, we assume that the variables are already sorted, i.e. $\mathcal{M}(\boldsymbol{x}_1) \geq \mathcal{M}(\boldsymbol{x}_2) \geq \ldots \geq \mathcal{M}(\boldsymbol{x}_D)$, where $\mathcal{M}(\boldsymbol{x}_j)$ denotes the number of missing cells in variable $\boldsymbol{x}_j$. Note that $\boldsymbol{x}_j$ denotes now the j-th column of the data matrix.

Step 3: Set $l = 1$.

Step 4: Use the ilr transformation (8.8) to transform the compositional data set.

Step 5: Denote $m_l \subset \{1, \ldots, n\}$ the indices of the observations that were originally missing in variable $\boldsymbol{x}_l$, and $o_l = \{1, \ldots, n\} \setminus m_l$ the indices corresponding to the observed cells of $\boldsymbol{x}_l$. Furthermore, $\boldsymbol{z}_l^{o_l}$ and $\boldsymbol{z}_l^{m_l}$ denote the l-th balance with the observed and missing parts, respectively, corresponding to the variable x_l. Let $Z_{-l}^{o_l}$ and $Z_{-l}^{m_l}$ denote the matrices with the remaining balances corresponding to the observed and missing cells of x_l, respectively. Additionally, the first column of $Z_{-l}^{o_l}$ and $Z_{-l}^{m_l}$ consists of ones, taking care of an intercept term in the regression problem

$$z_l^{o_l} = Z_{-l}^{o_l}\beta + \varepsilon \qquad (8.12)$$

with unknown regression coefficients β and an error term ε.

Step 6: Estimate the regression coefficients β in Equation (8.12), and use the estimated regression coefficients $\hat{\beta}$ to replace the missing parts $z_l^{m_l}$ by

$$\hat{z}_l^{m_l} = Z_{-l}^{m_l}\hat{\beta}. \qquad (8.13)$$

Step 7: Use the updated balances for back-transformation to the simplex with Equations (8.9)-(8.11). As a consequence, the values that were originally missing in the cells o_l in variable x_l are updated.

Step 8: Carry out Steps 4–7 in turn for each $l = 2, \ldots, D$.

Step 9: Repeat Steps 3–8 until the Euclidean distance between the empirical covariance matrices computed from the ilr transformed data according to Equation (8.8) from the present and the previous iteration is smaller than a certain boundary.

Although we have no proof of convergence, experiments with real and artificial data have shown that the algorithm usually converges in a few iterations, and that after the second iteration no significant improvement is obtained.

Note that the choice of the balances by Equation (8.8) is of advantage in Step 5 in the iterative procedure, because already for $l = 1$, the information of variable

x_1 with the highest amount of missings is only contained on the left hand side of (8.12), but not in the explanatory variables on the right hand side.

The estimation of the regression coefficients in Step 6 can be done in the classical way by using least-squares (LS) estimation. However, in presence of outliers we recommend using robust regression which is able to reduce the influence of outlying observations for estimating the regression parameters in Equation (8.12) [see, e.g., R. et al., 2006]. In our experiments we used least trimmed squares (LTS) regression because it is highly robust and fast to compute [Rousseeuw and Van Driessen, 2002]. Note that robust regression also protects against poorly initialized missing values, because the ilr transformation (8.8) can lead to contamination of all other cells in the corresponding observation.

8.4 Numerical study with a data example

The methods described in the previous section are applied to a data set used in Aitchison [1986], p. 395. This data set contains household expenditures on five commodity groups of 20 single men. The variables represent housing (including fuel and light), foodstuff, alcohol and tobacco, other goods (including clothing, footwear and durable goods) and services (including transport and vehicles). Thus they represent the ratios of the men's income spent on the mentioned expenditures. Since this data set is complete, we set the first observation and the third part to missing. The observed value in this cell is 147 HK$ (former Hong Kong dollar).

Additionally, we modify the third observation to see the influence of an outlier on the results of the imputation. Using the procedure of Filzmoser and Hron [2008a] one can see that this observation already represents an outlier in the data set. We multiplied the value in the third column by the factors 1 (i.e. the original data set), 2 and 10 to represent a person which is a possible alcoholic. This observation is then an outlier corresponding to both, the Euclidean and Aitchison geometry. This kind of outlier is denoted as *outlier 1* in the following. In a second outlier scenario, the third row of the original table was multiplied by the factors 1, 2 and 10 to obtain the second kind of outlier which is denoted as *outlier 2* in the following. Depending on the factor, the third person has in general more money to spend, but the proportions remain the same. Therefore this multiplication will not change anything in the Aitchison geometry, but it can affect the estimation in the Euclidean geometry (see Section 2).

The missing value is now estimated with various imputation techniques. The results are shown in Table 8.1. We apply methods that take the compositional nature of the data into account (geometric mean imputation, iterative LS in the ilr space, iterative LTS (ilr), imputation with the alr-EM algorithm, and $k$nn using the Aitchison distance). For comparison we also use methods that ignore the compositional nature of the data (arithmetic mean, EM algorithm, iterative LS without transformation, iterative LTS without transformation, $k$nn based on the Euclidean distances).

The resulting values in Table 8.1 demonstrate that methods that account for the compositional nature of the data lead to an improvement of the estimation. Focusing on the *outlier 1* scenario, the more extreme the outlier, the worse the results using standard methods without transformation and non-robust methods in the ilr space (geometric mean and LS (ilr)). For the *outlier 2* situation, the value of the factor does not change anything in the Aitchison geometry. Therefore, the iterative procedures based on LS and LTS regression in the ilr-space, $k$nn imputation based on Aitchison distances, and the alr-EM algorithm give the same results as for the original data. When working in the appropriate space, the model-based procedures are able to improve the initialized values from $k$nn imputation (for $k$nn we used $k = 4$ which gave the best results). The best result is obtained for LS (ilr) with factor 2 in scenario *outlier 1*. Although in this case the outlier has spoiled the regression hyperplane(s), the estimation has been improved by accident, because moving the outlier even further away (factor 10) leads to a severe underestimation.

Let us remark that by far the worst results are obtained for the univariate imputation methods arithmetic mean imputation and geometric mean imputation, although in the latter case the geometry on the simplex is taken into account. The standard EM algorithm produces more satisfactory results, but the outlier has a big influence on the quality of the estimation.

Although one cannot draw general conclusions from this simple numerical study, it gives a first impression about the performance of different imputation methods. In the next section we will consider more general situations using simulated data. Also a more detailed comparisons between the observed and the imputed values will be made based on the Aitchison distance and on the covariance structure.

8.5 Simulation study

For simulating compositional data we will use the so-called normal distribution on the simplex. A random composition x follows this distribution if, and only if, the vector of ilr transformed variables $z = ilr(x)$ follows a multivariate normal distribution on R^{D-1} with mean vector μ and covariance matrix Σ [see, e.g., Pawlowsky-Glahn et al., 2005, Mateu-Figueras and Pawlowsky-Glahn, 2008]. Thus, $x \sim \mathcal{N}_S^D(\mu, \Sigma)$ denotes that a D-part composition x is multivariate normally distributed on the simplex. Note that normality on the simplex is independent from the chosen balances for the ilr transformation.

In a first simulation study we generate $n = 100$ realizations from the random variable $x \sim \mathcal{N}_S^3(\mu, \Sigma)$ with

$$\mu = \begin{pmatrix} 0 \\ 2 \end{pmatrix} \quad \text{and} \quad \Sigma = \begin{pmatrix} 5.05 & 4.95 \\ 4.95 & 5.05 \end{pmatrix}.$$

Table 8.1: Estimations of the missing value in cell $[1,3]$ of the expenditures data set (observed value is 147). The considered imputation methods are geometric and arithmetic mean imputation (*gmean* and *mean*), iterative LS and LTS procedure with and without ilr transformation, EM algorithm for alr-transformed (*alr-EM*) and non-transformed (*EM*) data, and $k$nn imputation based on Aitchison and Euclidean distances. *original* corresponds to results for the original data, *outlier 1* is for an outlying observation in both the Aitchison and Euclidean geometries, *outlier 2* is for an outlier only in the Euclidean space. The numbers 1, 2, and 10 are the multiplication factors for generating the outliers.

observed value: 147	*original*	*outlier 1*		*outlier 2*	
method \ factor	1	2	10	2	10
gmean	289.6	300.3	326.9	289.6	289.6
alr-EM	157.8	155.4	150.1	157.8	157.8
$k$nn (Aitch.)	152.1	152.1	152.1	152.1	152.1
LS (ilr)	150.8	148.1	142.2	150.8	150.8
LTS (ilr)	150.8	150.3	150.3	150.8	150.8
mean	330.2	368.4	673.6	368.4	673.6
EM	190.2	214.9	798.4	163.5	195.6
$k$nn (Eucl.)	155.0	155.0	155.0	155.0	155.0
LS (no transf.)	161.0	179.2	324.5	161.3	160.3
LTS (no transf.)	161.3	158.6	158.6	161.4	153.6

Using the spectral decomposition, the matrix Σ can be re-written as

$$\Sigma = 5(1,1)\begin{pmatrix}1\\1\end{pmatrix} + 0.05(1,-1)\begin{pmatrix}1\\-1\end{pmatrix}.$$

Thus, the main variability of the generated data is along one direction, and the variability along the orthogonal direction is small. The resulting covariance matrix thus is poorly conditioned. Although this choice seems quite artificial, situations where only one out of many directions contains an essential portion of variability are realistic in compositional data sets [Kovács et al., 2006]. Typical examples are the Arctic lake sediment and Aphyric Skye lavas data sets [Aitchison, 1986, p. 359, 360]. Each observation is then multiplied by a factor generated from the uniform distribution $U(0,1)$ on the interval $(0,1)$. This multiplication does not change the equivalence class of the compositional data.

In order to see the influence of outliers on the different imputation methods, $n_1 + n_2$ out of the n observations will be replaced by the following types of outliers:

Outlier group 1: n_1 observations are simulated from $\mathcal{N}_S^3(\boldsymbol{\mu}_1, \Sigma)$, with $\boldsymbol{\mu}_1 = (0,6)^t$, and they are multiplied by a factor generated from $U(0,10)$.

Outlier group 2: n_2 observations are taken from the distribution $\mathcal{N}_S^3(\boldsymbol{\mu}, \Sigma)$, and they are multiplied by a factor coming from $U(0,10)$.

Within the simulation scheme the numbers of outliers in both groups will be the same, and they will be increased from 0 to 40, i.e. $n_1 = n_2 = 0, 1, \ldots, 40$.

The role of the outlier groups is similar to the two types of outliers in the numerical study of Section 4. *Outlier group 1* consists of observations that are potential outliers in the Euclidean space as well as in the Aitchison space, whereas *outlier group 2* will not have any effect in the Aitchison geometry due to the properties of equivalence classes for compositions (see Section 2).

The amount of missing values is fixed but different for each variable. 20% of the values in the first variable, and 10% in the second variable are set to be missing completely at random. Missing values are only generated in the non-outlying data group (also in the subsequent simulations), because here a fair comparison of different (classical and robust) methods is possible.

The advantage of this design in low dimension is that it is possible to visualize the generated three-part compositions in a ternary diagram. A ternary diagram is an equilateral triangle $X_1 X_2 X_3$ such that a composition $x = (x_1, x_2, x_3)^t$ is plotted at a distance x_1 from the opposite side of vertex X_1, at a distance x_2 from the opposite side of vertex X_2, and at a distance x_3 from the opposite side of vertex X_3 [see, e.g., Aitchison, 1986]. Figure 8.3 (left) shows a simulated data set where each of the outlier groups represents 5% of the observations. The 5 observations from *outlier group 1* are shown with symbol $+$, the 5 points from *outlier group 2* have symbol $\triangle$, and the remaining regular points are plotted with small dots. The points $+$ from *outlier group 1* are close to the boundary of the ternary diagram, and the points $\triangle$ from *outlier group 2* are spread over the main data cloud. Note that the points $\triangle$ do not appear as outliers in the ternary diagram because the different sums of the parts cannot be visualized. Figure 8.3 (right) shows the two dimensions of the ilr transformed data. Here the shifted center of the points $+$ from *outlier group 1* is clearly visible, whereas the points $\triangle$ are not acting as outliers in this space.

The choice of the parameters μ and Σ in the simulation study determines the shape of the raw (compositional) data on the simplex. The further away μ is from the null vector (indicates equilibrium on the simplex), the closer the compositions are to the border of the simplex. The choice of the covariance matrix Σ determines how elongated the data points appear in the ternary diagram.

The original and the imputed data values are compared by two different criteria:

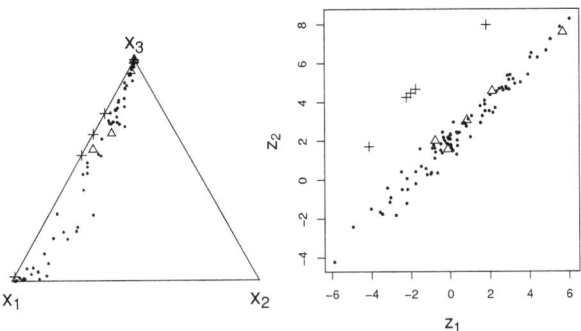

Figure 8.3: Simulated data set with 5 points from *outlier group 1* (symbol +) and 5 points from *outlier group 2* (symbol △). Left plot: 3-part compositions plotted in the ternary diagram; right plot: data after ilr transformation.

Compositional error variance: Let $M \subset \{1, \ldots, n\}$ denote the index set referring to observations including at least one missing cell, and $n_M = |M|$ be the number of such observations. The *compositional error variance* is defined as

$$\frac{1}{n_M} \sum_{i \in M} d_A^2(\boldsymbol{x}_i, \hat{\boldsymbol{x}}_i) \tag{8.14}$$

where $\boldsymbol{x}_i$ denotes the original composition (before setting cells to missing), and $\hat{\boldsymbol{x}}_i$ denotes the composition where only the missing cells are imputed.

Difference in covariance structure: We denote $\boldsymbol{S} = [s_{ij}]$ as the sample covariance matrix of the non-outlying original ilr transformed observations z_{ij}, using for example the transformation from Equation (8.8) (another ilr basis would not change our criterion introduced in the following). Further, $\tilde{\boldsymbol{S}} = [\tilde{s}_{ij}]$ denotes the sample covariance matrix computed with the same ilr transformed observations where all missing cells have been imputed. The *difference in covariance structure* is based on the Euclidean distance between both covariance estimations, namely as

$$\frac{1}{D-1}\sqrt{\sum_{i=1}^{D-1}\sum_{j=1}^{D-1}(s_{ij}-\tilde{s}_{ij})^2} = \frac{1}{D-1}\|\boldsymbol{S}-\tilde{\boldsymbol{S}}\|. \quad (8.15)$$

Thus the *compositional error variance* measures closeness of the imputed values in the Aitchison geometry, whereas the influence of the imputation to the multivariate data structure is expressed by the *difference in covariance structure*.

For each considered percentage of outliers (0% to 40% per outlier group, in steps of 1%) we simulated 1000 data sets according to the above scheme, and estimated the missing values with different techniques. Then the average for the above quality criteria is computed. The results are displayed in Figure 8.4, and they confirm the findings of the numerical study in Section 4. The left column of the pictures in Figure 8.4 shows the results for the non-transformed data (the imputation is done in the Euclidean space), the right column is for the transformed data (the imputation is done in the Aitchison geometry). The top row of the pictures shows the average of the *compositional error variance*, the bottom row presents the average *difference in covariance structure*. Since we use the same scale, it is easy to compare the pictures of one row. If the compositional nature of the data is ignored (left column), the results generally get worse. An exception are the $k$nn results for the *difference in covariance structure*, which are similar when using the Euclidean or the Aitchison distance. For kNN the best results were achieved for $k = 8$, but other choices have a rather small effect on the outcome.

When working in the proper geometry (right column of the pictures in Figure 8.4), the initial kNN imputation can be improved considerably with the iterative model-based imputation. If no outliers are present, the use of LS-regressions or LTS-regressions within the model-based procedure leads to comparable results. However, in presence of outliers the robust imputation algorithm shows clear ad-

CHAPTER 8. IMPUTATION OF MISSING VALUES FOR CODA

8.5. SIMULATION STUDY

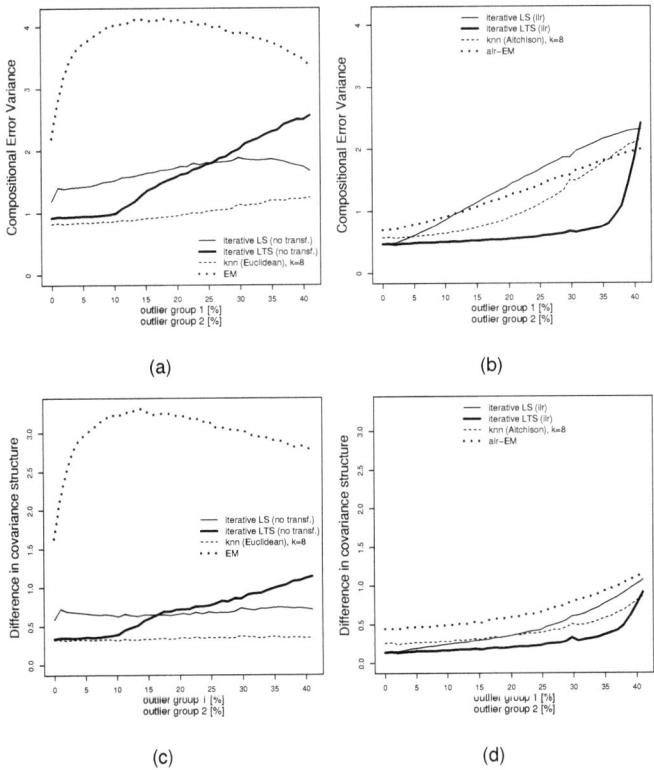

Figure 8.4: Simulation results (average of the *compositional error variance* and the *difference in variation*) for kNN imputation, model-based imputation using iterative LS and LTS regressions and EM/alr-EM algorithm. For (a) and (c) the imputation is done in the Euclidean space (no transformation), for (b) and (d) the imputation methods are applied in the Aitchison geometry.

vantages over LS. The iterative LTS procedure remains very stable up to about 35% outliers.

Since the previous simulation study is limited to low dimensionality and to a special choice of the covariance matrix, we want to provide deeper insight with more general situations. Thus, in a second and third simulation setup $n = 100$ realizations from the random variable $x \sim \mathcal{N}_S^{10}(\mu, \Sigma)$ are generated (higher dimension). μ is chosen as the null vector (equilibrium on the simplex), and all diagonal elements of Σ are equal to 1. The off-diagonal elements of Σ are chosen as 0.9 for the second setup, and 0.1 for the third setup. We thus can expect that the second setup will be advantageous for the model-based procedures, but that the third setup is particularly difficult because of the low correlations. Only outliers of type 1 are considered, and they are sampled from $\mathcal{N}_S^{10}(\mu_1, \Sigma)$, with $\mu_1 = (6, 0, \ldots, 0)^t$. The missing values are generated with MCAR. For the second setup, the percentages of missing values in variables 1 to 10 were chosen as 20%, 10%, 5%, 2%, ..., 2%, 0%, respectively. In the third setup they are taken as 5%, only the last variable is without missings (which is important for the alr-EM algorithm).

Figure 8.5 shows the results from 1000 simulated data sets corresponding to the second (left column) and the third (right column) simulation configuration. The results of only those imputation methods are shown which take the compositional nature of the data into account; results of the other methods are generally worse, although no outlier group 2 is considered. For the second setup (Figure 8.5, (a) and (c)), the iterative LS method yields the best results in the uncontaminated case. In presence of up to 30% outliers, the iterative LTS method shows the best performance. Outliers have a large effect on all other methods to both, the compositional error variance and the difference in covariance structure. The most difficult situation, the third simulation setup, leads to a similar behavior of all considered imputation methods (Figure 8.5, right column). In presence of outliers the iterative LTS method performs slightly better. 5model-based imputation using iterative LS and LTS regressions and alr-EM. (a)

Finally, in our simulation the missings were generated according to an MCAR situation. We also tested the MAR situation, leading to analogous conclusions. The reason is that multivariate imputation methods such as kNN and the proposed model-based approach can deal with MAR, but univariate mean imputation techniques which have shown very poor performance already for MCAR cannot deal with the MAR situation either.

We also investigated other imputation methods in the Euclidean and in the Aitchison geometry, such as univariate approaches based on arithmetic and geometric mean, the EM algorithm [Dempster et al., 1977], iterative procedures based on principal component analysis (PCA) [Serneels and Verdonck, 2008] and their robust counterparts [Fritz and Filzmoser, 2008], Bayesian PCA [S. et al., 2003], and probabilistic PCA [Bishop, 1999]. Also various other well-known imputation methods for which an implementation in R [R Development Core Team, 2008a] is available were tested [Troyanskaya et al., 2001, Kim et al., 2005, Scholz et al., 2005] (additional methods would be available in R, but sometimes the code is erroneous). All these methods give worse results, and in order to avoid confusion on the plots they are not shown in Figure 8.4 and 8.5 . In addition, we tested several variants for kNN imputation (as listed in Section 8.3.1), in combination with different measures of location (arithmetic mean, median) for the aggregation of the k-nearest neighbors, leading to poorer performance.

8.6 Conclusions

In general, the estimation of missing values in multivariate data can be done more reliably with multivariate rather than with univariate imputation techniques.

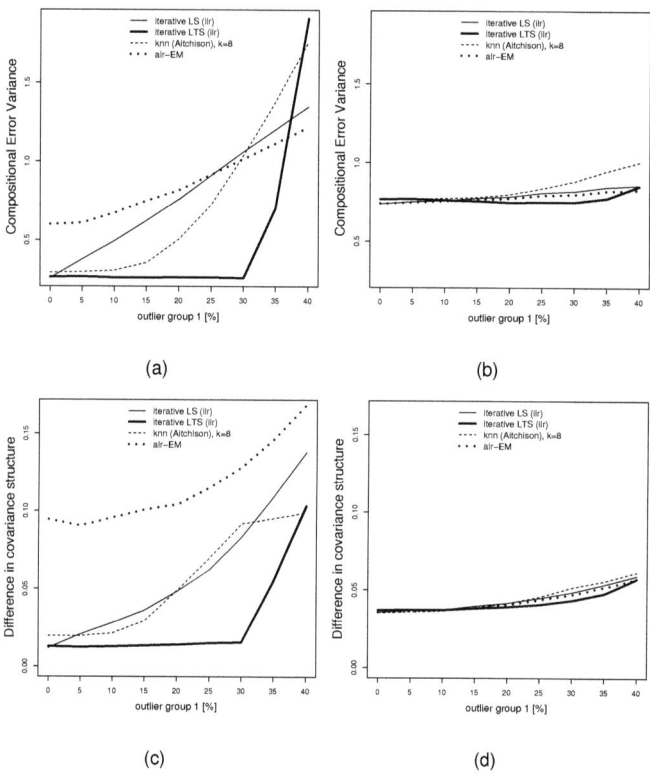

Figure 8.5: Simulation results (average of the *compositional error variance* and the *difference in variation*) for kNN imputation, model-based imputation using iterative LS and LTS regressions and alr-EM. (a) and (c) shows the results from the second setup, (b) and (d) presents those from the third setup.

However, such procedures are depending on a realistic estimation of the multivariate data structure, because they are either model-based, covariance-based, distance-based, or use a combination of these approaches. The estimation of either the covariance matrix or the distance matrix is sensitive to outliers or data inhomogeneities. Even worse, the underlying geometry of the data plays an essential role for the estimation. In the context of compositional data one has to account for the fact that only relative information in form of ratios between the variables is relevant. In other words, one has to work in the Aitchison geometry rather than in the usual Euclidean geometry [Aitchison, 1986].

We have proposed two imputation methods for estimating missing values in compositional data. The first method uses the information of the k-nearest neighbors, being defined via the Aitchison distance [Aitchison et al., 2000]. In principle, this method is not robust against outliers, since an outlying cell in an observation will change the distance. On the other hand, if there are other observations without outlying cells that include valuable information for imputation, the estimation of the missing value might not get much worse. Moreover, one can increase k for kNN imputation in order to collect enough reliable information from the neighbors. This relative robustness of kNN imputation is also visible in the simulation results (see Figure 8.4 and 8.5). In presence of contamination the robust adjustment according to Equation (8.7) led to better results than the adjustment according to Equation (8.5) in all our tests.

The second proposed method uses kNN imputation to initialize the missing cells. Then regressions are performed in an iterative manner, where in each step one variable serves as the response variable, and the remaining variables as the regressors. This scheme allows to estimate the missing cells in the response with the multivariate information contained in the regressors. However, the regressions have to be done in the appropriate geometry, and therefore we switch to the ilr space [Egozcue et al., 2003] with a special choice of the ilr basis. For homogeneous, outlier-free data one can use least-squares regressions, but in presence of outliers we recommend using robust regressions, like LTS regression [Rousseeuw and Van Driessen, 2002]. The numerical study (Section 4) and the simulations (Section 5) have demonstrated that the initial kNN estimation can be essentially improved by this approach. The simulation study has shown that not only the distances between the true and the estimated missings remain very stable, even in presence of outliers, but also the multivariate data structure is well preserved. Our model-based imputation procedure outperformed also other multivariate imputation techniques, even if they are applied in the appropriate geometry.

The simulation was carried out for various different parameter configurations. Three of these simulation setups are presented in this paper whereas one configuration (the third) could be considered as the worst case, especially for our proposed iterative model based imputation technique, because the correlations between the parts are very low, the amount of missing values is equal in each variable, and moderate outliers were generated only in one dimension. Nevertheless, all our results lead to the conclusion that the proposed iterative method has comparable performance to other imputation methods in the worst case, but outperforms existing imputation methods in less extreme situations.

An implementation of our proposed procedures in R [R Development Core Team, 2008a] is available in the package *robCompositions* at the Comprehensive R Archive Network, see http://cran.r-project.org/.

9 Compositional Data Using the R-Package robCompositions

Published in the Proceedings of the NTTS (New Technologies and Techniques in Statistics) conference 2009 [Templ et al., 2009a]

Matthias Templ**,***, Peter Filzmoser***, Karel Hron*,

* Department of Methodology, Statistics Austria, Guglgasse 13, 1110 Vienna, Austria. (matthias.templ@statistik.gv.at) and

** Department of Statistics and Probability Theory, Vienna University of Technology, Wiedner Hauptstr. 8-10, 1040 Vienna, Austria. (templ@statistik.tuwien.ac.at)

*** Department of Mathematical Analysis and Applications of Mathematics, Palacký University, Tomkova 40, 779 00 Olomouc, Czech Republic

Abstract: The aim of this contribution is to show how the R-package robCompositions can be applied to estimate missing values in compositional data. Two procedures are summarized, one of them being highly stable also in presence of outlying observations. Measures for information loss are presented, and it is demonstrated how they can be applied. Moreover, we introduce new diagnostic tools that are useful for inspecting the quality of the imputed data.

9.1 Introduction

9.1.1 Imputation

Many different methods for imputation have been developed over the last few decades. The techniques for imputation can be subdivided into four categories: univariate methods such as column-wise (conditional) mean or median imputation, distance-based imputation methods such as k-nearest neighbor imputation, covariance-based methods such as the well-known expectation maximization imputation method, and model-based methods such as regression imputation. Most of these methods are able to deal with missing completely at random (MCAR) and missing at random (MAR) missing values mechanism (see, e.g. Little and Rubin [1987]). However, most of the existing methods assume that the data originate from a multivariate normal distribution. This assumption becomes invalid as soon as there are outliers in the data. In that case imputation methods based on robust estimates should be used.

9.1.2 Compositional Data

Advanced (robust) imputation methods have turned out to work well for data with a direct representation in the Euclidean space. However, this is not the case when dealing with compositional data.

An observation $x = (x_1, \ldots, x_D)$ is called a D-part composition if, and only if, all its components are strictly positive real numbers and all the relevant information is included in the ratios between them [Aitchison, 1986]. One can thus define the *simplex*, which is the sample space of D-part compositions, as

$$\mathcal{S}^D = \{x = (x_1, \ldots, x_D), x_i > 0, \sum_{i=1}^{D} x_i = \kappa\} \ . \tag{9.1}$$

Note that the constant sum constraint κ implies that D-part compositions are only $D-1$ dimensional, so they are singular by definition. It is, however, possible that

9.1. INTRODUCTION

the constant κ is different for each observation (for further details, see Hron et al. [2008a]). In any case, the important property of compositional data is that all information is contained in the ratios of the parts.

The application of standard statistical methods, like correlation analysis or principal component analysis, directly to compositional data can lead to biased and meaningless results [Filzmoser and Hron, 2008a,b]. This is also true for imputation methods [Bren et al., 2008, Martín-Fernández et al., 2003, Boogaart et al., 2006]. A way out is to first transform the data with appropriate transformation methods. Such transformations, preserving the specific geometry of compositional data on the simplex (also called Aitchison geometry), are represented by the family of log-ratio transformations: additive, centered [Aitchison, 1986] and isometric (abbreviated by *ilr*, [Egozcue et al., 2003] transformations. Standard statistical methods can then be applied to the transformed data, and the results can be back-transformed.

Compositional data frequently occur in official statistics. Examples are expenditure data, income components in tax data, wage components in the Earnings Structure Survey, components of turnover of enterprises etc., and all data which sum up to a constant or which carry all the information only in the ratios. The problem of missing values in compositional data including outliers is a common problem not only in official statistics, but also in various other fields (see, e.g., [Graf, 2006, Filzmoser and Hron, 2008a]). In the following Section we will briefly review two algorithms for imputation that are described in detail in Hron et al. [2008a]. Section 3 focuses on the use of the R-package robCompositions, and Section 4 introduces some diagnostic tools implemented in this package. The final Section 5 concludes.

9.2 Proposed Imputation Algorithms

In the following we briefly describe the imputation methods that have been implemented in the R-package `robCompositions`. The detailed description of the algorithms can be found in Hron et al. [2008a].

9.2.1 k-Nearest Neighbor Imputation

k-nearest neighbor imputation usually uses the Euclidean distance measure. Since compositional data are represented only in the simplex sample space, we have to use a different distance measure, like the Aitchison distance, being defined for two compositions $\boldsymbol{x} = (x_1, \ldots, x_D)$ and $\boldsymbol{y} = (y_1, \ldots, y_D)$ as

$$d_a(\boldsymbol{x}, \boldsymbol{y}) = \sqrt{\frac{1}{D} \sum_{i=1}^{D-1} \sum_{j=i+1}^{D} \left(\ln \frac{x_i}{x_j} - \ln \frac{y_i}{y_j} \right)^2}. \tag{9.2}$$

Thus, the Aitchison distance takes care of the property that compositional data include their information only in the ratios between the parts.

Once the k-nearest neighbors to an observation with missing parts have been identified, their information is used to estimate the missings. For reasons of robustness, the estimation is based on using medians rather than means. If the compositional data do not sum up to a constant, it is important to use an adjustment according the sum of all parts prior to imputation. For details, see Hron et al. [2008a].

9.2.2 Iterative Model-Based Imputation

In the second approach we initialize the missing values with the proposed k-nearest neighbor approach. Then the data are transformed to the $D - 1$ dimensional real space using the ilr transformation. Let d_e denote the Euclidean distance. The ilr transformation holds the so-called isometric property,

$$d_a(\boldsymbol{x}, \boldsymbol{y}) = d_e(ilr(\boldsymbol{x}), ilr(\boldsymbol{y})) \tag{9.3}$$

[Egozcue and Pawlowsky-Glahn, 2005]. Consequently, one can use standard statistical methods like multiple linear regression, that work correctly in the Euclidean space.

We take a special form of the ilr transformation, namely $ilr(\boldsymbol{x}) = (z_1, \ldots, z_{D-1})$, with

$$z_i = \sqrt{\frac{D-i}{D-i+1}} \ln \frac{\sqrt[D-i]{\prod_{j=i+1}^{D} x_j}}{x_i} \quad \text{for } i = 1, \ldots, D-1 \ . \quad (9.4)$$

Here, the compositional part x_1 includes the highest amount of missings, x_2 the next highest, and so on. Thus, when performing a regression of z_1 on $z_2, \ldots, z_{D-1}$, only z_1 will be influenced by the initialized missings in x_1, but not the remaining ilr variables.

The idea of the procedure is thus to iteratively improve the estimation of the missing values. After the regression of z_1 on $z_2, \ldots, z_{D-1}$, the results are back-transformed to the simplex, and the cells that were originally missing are updated. Next we consider the variable which originally has the second highest amount of missings, and the same regression procedure as before is applied in the ilr space. After each variable containing missings has been proceeded, one can start the whole process again until the estimated missings stabilize. The detailed description of this algorithm can be found in Hron et al. [2008a].

As a regression method we propose to use robust regression, like LTS regression (see R. et al. [2006]), especially if outliers might be present in the data.

9.3 Using the R-package robCompositions for Imputing Missing Values

9.3.1 Data

The package includes the three compositional data sets *aitchison359*, *aitchison360*, and *aitchison395*, that have been published in Aitchison [1986]. In the following, however, we will use simulated data, where the data structure and outliers are exactly known. The data generation is the same as described in Hron et al. [2008a], and a plot of the data set in shown in Figure 9.1 for the original data (left) and for the ilr transformed data (right): We took 90 observations with 3 parts that are normally distributed on the simplex (i.e. they are multivariate normally distributed in the 2-dimensional ilr space). A group of 5 outliers (*group 1*) is added (green crosses in Figure 9.1) that are potential outliers in the Aitchison and in the Euclidean space. Another group (*group 2*) of 5 outliers (blue triangles in Figure 9.1) is only affecting the Euclidean space. Note that both types of outliers are simulated to have a considerably higher sum of their parts, which is not visible in the ternary diagram [Aitchison, 1986] in Figure 9.1 (left) where the parts are re-scaled to have sum 1.

The generated (complete) data are stored in the list element z2 of object x. Among the non-outliers we set 20% of the values in the first part, 10% in the second, and 5% in the third part to missing, using an MCAR mechanism. The new data set is stored in the list element zmiss of object x.

9.3.2 Usage of the Imputation Methods Within the Package

We apply k-nearest neighbor imputation for the generated compositional data set, and use the parameter $k = 6$:

```
> library(robCompositions)
> xImp <- impKNNa(x$zmiss, k = 6)
```

As a default, Aitchison distances are used for identifying the k-nearest neighbors (further options are provided, see help file). By default, the median is taken for re-scaling the k-nearest neighbors for imputation, but also other choices are possible.

The resulting object xImp is of class

CHAPTER 9. COMPOSITIONAL DATA USING THE R-PACKAGE ROBCOMPOSITIONS
9.3. USING THE R-PACKAGE ROBCOMPOSITIONS FOR IMPUTING MISSING VALUES

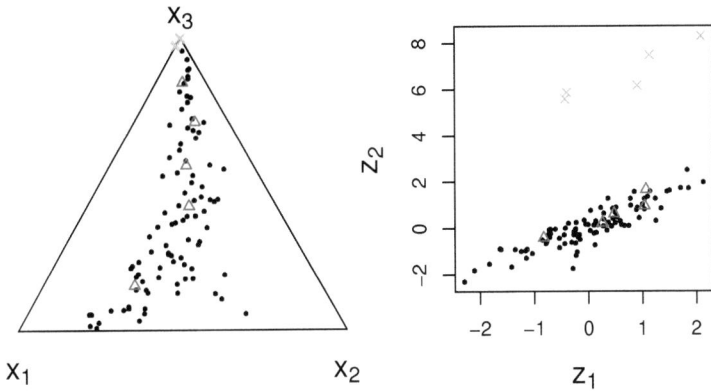

Figure 9.1: Simulated data set with 5 points from *outlier group 1* (symbol) and 5 points from *outlier group 2* (symbol △). Left plot: 3-part compositions shown in the ternary diagram; right plot: data after ilr transformation.

```
> class(xImp)
[1] "imp"
```

A print, a summary, and a plot method are provided for objects of this class:

```
> methods(class = "imp")
[1] plot.imp     print.imp    summary.imp
> xImp

[1] "31 missing vales were imputed"
```

Various informations are included in the object xImp, which can be accessed easily:

```
> names(xImp)
[1] "xOrig"    "xImp"      "criteria"  "iter"     "w"
```

```
"wind"  "metric"
```

The list element xOrig contains the original data, xImp is the imputed data set, w contains the number of missing values, and wind includes the indices of the missing values (imputed values). All this information is needed in order to provide suitable summaries and diagnostic plots.

The iterative model-based imputation method is applied with:

```
> xImp1 ← impCoda(x$zmiss, method = "lm")
> xImp2 ← impCoda(x$zmiss, method = "ltsReg")
```

The first command uses classical least-squares regression within the algorithm, the second command takes robust LTS regression.

9.4 Information Loss, Uncertainty, and Diagnostics

The quality of the imputed values can be judged by different criteria. We can use information loss criteria and compute the differences of the imputed to the observed data. If the observed data are known, we can use the bootstrap technique for measuring the uncertainty of the imputation. If the observed data are not known, diagnostic plots can be used for visualizing the imputed values.

9.4.1 Information Loss Measures

We compare the imputed and the original data values by two different criteria:

Relative Aitchison distance: **(RDA)** Let $M \subset \{1, \ldots, n\}$ denote the index set referring to observations that include at least one missing cell, and $n_M = |M|$ be the number of such observations. We define the *relative Aitchison distance* as

$$\frac{1}{n_M} \sum_{i \in M} d_A(\boldsymbol{x}_i, \hat{\boldsymbol{x}}_i) \tag{9.5}$$

where x_i denotes the original composition (before setting cells to missing), and $\hat{x}_i$ denotes the composition where only the missing cells are imputed.

Difference in variations: **(DV)** We use the variation matrix $T = [t_{ij}]$, with

$$t_{ij} = \text{var}\left(\ln \frac{x_i}{x_j}\right), \quad i,j = 1,\ldots,D,$$

and the empirical variance for var. Thus, t_{ij} represents the variance of the log-ratio of the parts i and j. Here, only the non-outlying original observations are considered for computing T. On the other hand, $\tilde{T} = [\tilde{t}_{ij}]$ denotes the variation matrix computed for the same observations, where all missing cells have been imputed. Then we define the *difference in variations* as

$$\frac{2}{D(D-1)} \sum_{i=1}^{D-1} \sum_{j=i+1}^{D} |t_{ij} - \tilde{t}_{ij}| \tag{9.6}$$

Thus, RDA measures closeness of the imputed values in the Aitchison geometry, whereas the influence of the imputation to the multivariate data structure is expressed by DV.

Using the iterative model-based algorithm for our test data set, we can show that the robust procedure based on LTS regression gives more reasonable results than its classical counterpart (the code for computing the measures is snipped):

```
[1] "RDA: iterative lm approach: 0.529"
```

```
[1] "RDA: iterative ltsReg approach: 0.307"
```

```
[1] "DV: iterative lm approach: 0.052"
```

```
[1] "DV: iterative ltsReg approach: 0.007"
```

9.4.2 Measuring the Uncertainty of the Imputations

Little and Rubin [1987] suggests to estimate standard errors for estimators via bootstrapping, and he outlines two approaches - a modified bootstrap approach and a modified jackknife procedure - to measure consistent standard errors when data will be imputed.

We draw bootstrap samples from both, the original data without missings, and the data where some values were set to missing, hereby using the same random seeds. For the latter bootstrap samples we impute the missing values with mean imputation (column-wise arithmetic mean), and classical and robust iterative model-based imputation. We are interested in the geometric mean of each variable. Figure 9.2 shows boxplots of the resulting geometric means (computed only for the non-outlying observations) for $r = 1000$ bootstrap replicates. The red horizontal lines indicate the geometric means for the original data without outliers.

It is clearly visible that mean imputation - a simple method still frequently applied - can lead to higher uncertainty, and that the results are biased. The model-based procedures have a very similar behavior as the original data.

9.4.3 Diagnostic Plots

Here we do not assume knowledge about the observed values. The goal is to visualize the imputed values in an appropriate way. Because of space limitations we only show results for the robust model-based procedure.

The first diagnostic plot is a multiple scatterplot where the imputed values are highlighted. The plot can be generated with `plot(xImp2, which=1)`. Figure 9.3 shows the result, and we can see that the imputed values are placed on the regression hyperplane(s). If this should be avoided, one can add random noise to the imputed values. This can be done with the function `impCoda()` using the parameter `method = ltsReg2` which considers the standard deviation of the residu-

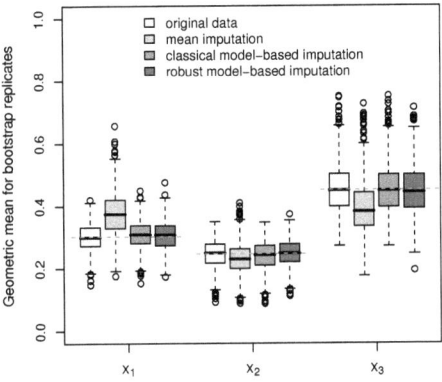

Figure 9.2: Boxplot comparison of the estimated column-wise geometric means of 1000 bootstrap replicates of the original data set and the data sets where missing values were imputated with different methods. The red line is the column-wise geometric mean of the original data. Outliers are excluded in the computation of the geometric means.

als for generating the random noise. This plot can also be generated for instance for log-ratio transformed data [Aitchison, 1986], taking care of the compositional nature of the data.

A parallel coordinate plot [Wegman, 1990] can be generated by plot(xImp2, which=2) (plot not shown here). The imputed values in certain variables are highlighted. One can select variables interactively, and imputed values in any of the selected variables will be highlighted.

The third diagnostic plot (see Figure 9.4), a ternary diagram [Aitchison, 1986], can be generated by plot(xImp2, which=3, seg1=FALSE). The 3-part compositions are presented by three spikes, pointing in the directions of the corresponding three variables. The spikes of the imputed values are highlighted. This presen-

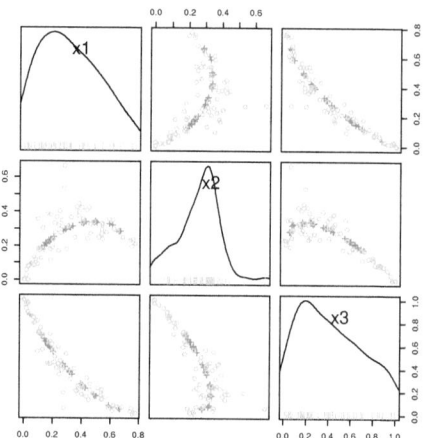

Figure 9.3: Multiple Scatterplot to highlight imputed observations.

tation allows gaining a multivariate view of the data, being helpful for interpreting possible irregularities of imputed values.

9.5 Conclusions

We provide the R-package robCompositions which includes advanced methods for imputation for compositional data. We have shown how the imputation methods described in Hron et al. [2008a] can be applied with the package. The methods are especially designed for data including outliers. The performance of the methods is outlined in the original paper.

The package includes possibilities for evaluating the quality of the imputed values: One can compute measures for information loss, use bootstrapping for estimating bias and uncertainty of parameters, and visualize the imputed values with diagnostic tools. For 3-dimensional compositions the proposed ternary plot is designed to highlight how well imputations are made and which compositions were

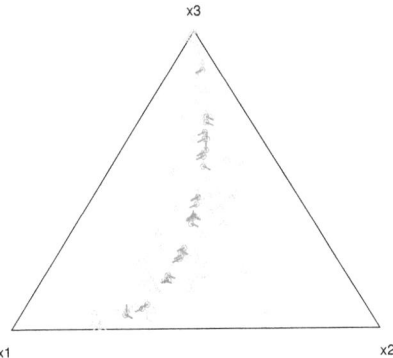

Figure 9.4: Ternary diagram with special plotting symbols for highlighting imputed parts of the compositional data.

imputed. Note that many additional options can be used within these plots given by the arguments of the plotting function.

Bibliography

E. Acuna and members of the CASTLE group. *dprep: Data preprocessing and visualization functions for classification*, 2008. URL http://math.uprm.edu/~edgar/dprep.html. R package version 2.0.

J. Aitchison. *The Statistical Analysis of Compositional Data*. Chapman & Hall, London, 1986.

J. Aitchison, C. Barceló-Vidal, J.A. Martín-Fernández, and V. Pawlowsky-Glahn. Logratio analysis and compositional distance. *Mathematical Geology*, 32(3): 271–275, 2000.

N. Anwar. Micro-aggregation - the small aggregates method. In *Internal report*. Luxembourg: Eurostat, 1993.

J. Bacher and S. Brand, R. andBender. Re-identifying register data by survey data using cluster analysis: An empirical study. *International Journal of Uncertainty, Fuzziness and Knowledge-Based Systems*, 10(5):589–608, 2002.

C. Beguin and B. Hulliger. The BACON-EEM algorithm for multivariate outlier detection in incomplete survey data. *Survey Methodology*, 34(1):91–103, 2008.

R. Benedetti and L. Franconi. Statistical and technological solutions for controlled data dissemination. In *Pre-Proceedings of New Techniques and Technologies for Statistics*, pages 225–232, 1998.

M. Berkelaar, J. Dirks, K. Eikland, and P. Notebaert. lpsolve ide v5.5, 2006.

J.G. Bethlehem, W.J. Keller, and J. Pannekoek. Disclosure control of microdata. *Journal of the American Statistical Association*, 85(409):38–45, 1990.

C.M. Bishop. Probabilistic principal component analysis. *Journal of the Royal Statistical Society*, 61(B):611–622, 1999.

K.G. Boogaart, R. Tolosana-Delgado, and M. Bren. Concept for handling with zeros and missing values in compositional data. In E. Pirard, editor, *CProceedings of IAMG'06 - The XI annual conference of the International Association for Mathematical Geology*. University of Liege, Belgium. CD-ROM., 2006. 4 pages.

L. Borchsenius. New developements in the danish system for access to micro data. In *Monographs of official statistics, Work session on statistical data confidentiality*. Eurostat, Luxembourg, 2005.

G.E.P. Box and D.R. Cox. An analysis of transformations. *Journal of the Royal Statistical Society*, Series B, 26:211–252, 1964.

R. Brand. Microdata protection through noise addition. In *Privacy in Statistical Databases. Lecture Notes in Computer Science. Springer*, pages 347–359, 2004.

R. Brand and S. Giessing. Report on preparation of the data set and improvements on sullivans algorithm. Technical report, CASC Project Deliverable 1.1-D1, 2002.

M. Brandt and H. Hafner. Leitfaden zur Anonymisierung für die Erstellung eines Campus-Files aus den Einzeldaten der zweiten europäischen Erhebung zur beruflichen Weiterbildung. Technical report, Statistisches Bundesamt, Hessisches Statistisches Landesamt., 2007. URL http://www.forschungsdatennetzwerk.de/bestand/cvts/cf/2000/fdz_cvts_cf_2000_leitfaden_zur_anonymisierung.pdf. Nr. 5, 10/2005.

M. Bren, R. Tolosana-Delgado, and K.G. van den Boogaart. News from compositions, the R package. In D. Estadella, J. Martín-Fernández, and J. Antoni, editors, *CoDaWork'08*. Universitat de Girona. Departament d'Informática i Matemática Aplicada, 2008.

J. Burridge. Information preserving statistical obfuscation. *Statistics and Computing*, 13:321–ï£¡327, 2003.

A. Capobianchi, S. Polettini, and M. Lucarelli. Strategy for the implementation of individual risk methodology into μ-ARGUS. Technical report, Report for the CASC project. No: 1.2-D1, 2001.

J. Castro. Network flows heutistics for complementary cell suppression. In *Lecture Notes in Computer Sciences. Inference Control in Statistical Databases.*, volume 2316, pages 59–73, 2002.

J. Castro and D. Baena. Using a mathematical programming modelling language for optimal CTA. *Privacy in Statistical Databases. Lecture Notes in Computer Science*. Springer, 5262:1–12, 2008. ISBN 978-3-540-87470-6.

J.M. Chambers. *Programming with Data*. Springer, New York, 1998. ISBN 0-387-98503-4.

D. Cook and D.F. Swayne. *Interactive and Dynamic Graphics for Data Analysis: With R and GGobi*. Springer, New York, 2007. ISBN: 978-0-387-71761-6.

L. Cox. Network models for complementary cell suppression. *Journal of the American Statistical Association*, 90:1453–1462, 1995.

L.H. Cox. Linear sensitivity measures in statistical disclosure control. *Journal of Statistical Planning and Inference*, 75:153–164, 1981.

C. Croux and A. Ruiz-Gazen. High breakdown estimators for principal components: the projection-pursuit approach revisited. *Journal of Multivariate Analysis*, 95:206–226, 2005.

T. Dalenius and S.P. Reiss. Data-swapping: A technique for disclosure control. In *Proceedings of the Section on Survey Research Methods*, volume 6, pages 73–85. American Statistical Association, 1982.

P.P. De Wolf. HiTaS: A heuristic approach to cell suppression in hierarchical tables. In *Lecture Notes in Computer Sciences. Inference Control in Statistical Databases*, volume 2316, pages 81–98, 2002.

D. Defays and Anwar M.N. Masking microdata using micro-aggregation. *Journal of Official Statistics*, 14(4):449–461, 1998.

D. Defays and P. Nanopoulos. Panels of enterprises and confidentiality: the small aggregates method. In *Proceedings of the 1992 Symposium on Design and Analysis of Longitudinal Surveys*, pages 195–204. Statistics Canada, Ottawa, 1993.

A.P. Dempster, N.M. Laird, and D.B. Rubin. Maximum likelihood for incomplete data via the EM algorithm (with discussions). *Journal of the Royal Statistical Society*, 39:1–38, 1977.

G. Dinges and M. (contributed equally) Templ. Motivation zur statistik - computergestützt lernen in der statistik austria. *Austrian Journal of Statistics*, 38(1): 1–16, 2009. The authors contributed equally to this work.

J. Domingo-Ferrer and J.M. Mateo-Sanz. Practical data-oriented microaggregation for statistical disclosure control. *IEEE Trans. on Knowledge and Data Engineering*, 14(1):189–201, 2002.

J. Domingo-Ferrer and V. Torra. A quantitative comparison of disclosure control methods for microdata. In *Confidentiality, Disclosure and Data Access: Theory and Practical Applications for Statistical Agencies*, pages 111–134, 2001.

J. Domingo-Ferrer, J.M. Mateo-Sanz, and V. Torra. Comparing sdc methods for microdata on the basis of information loss and disclosure risk. In *Pre-Proccedings of ETK-NTTS*, volume 2, pages 807–826. Springer, 2001.

J. Domingo-Ferrer, J.M. Mateo-Sanz, A. Oganian, and A. Torres. On the security of microaggregation with individual ranking: analytical attacks. *International Journal of Uncertainty, Fuzziness and Knowledge-Based Systems*, 10(5):477–492, 2002.

G.T. Duncan, S.A. Keller-McNulty, and S.L. Stokes. Disclosure risk vs. data utility: the r-u confidentiality map. Technical report la-ur-01-6428, Los Alamos National Laboratory, 2001. URL http://www.heinz.cmu.edu/research/122full.pdf.

C. Eaton, C. Plaisant, and T. Drizd. Visualizing missing data: Graph interpretation user study. In *Human-Computer Interaction - INTERACT 2005, Lecture Notes in Computer Sciences, Springer*, pages 861–872, 2005. ISBN 978-3-540-28943-2.

R.G. Efron and R.G. Tibshirani. *An Introduction to the Bootstrap*. Chapman and Hall, New York, 1993.

J.J. Egozcue and V. Pawlowsky-Glahn. Groups of parts and their balances in compositional data analysis. *Mathematical Geology*, 37(7):795–828, 2005.

J.J. Egozcue and V. Pawlowsky-Glahn. *Compositional data analysis in the geosciences: From theory to practice*, chapter Simplicial geometry for compositional data, pages 145–160. Geological Society, London, 2006. Special Publications 264.

J.J. Egozcue, V. Pawlowsky-Glahn, G. Mateu-Figueras, and C. Barceló-Vidal. Isometric log-ratio transformations for compositional data analysis. *Mathematical Geology*, 35(3):279–300, 2003. doi:10.1023/A:1023818214614. URL http://www.springerlink.com/content/wx1166n56n685v82/.

E.A.H. Elamir and C.J. Skinner. Record level measures of disclosure risk for survey microdata. *Journal of Official Statistics (Submitted)*, 2006.

M. Elliot, A. Hundepool, E.S. Nordholt, J-L. Tambay, and T. Wende. Glossary on statistical disclosure control, 2005. URL http://neon.vb.cbs.nl/casc/Glossary.htm.

P. Filzmoser. A multivariate outlier detection method. In S. Aivazian, P. Filzmoser, and Y. Kharin, editors, *Proceedings of the Seventh International Conference on Computer Data Analysis and Modeling*, volume 1, pages 18–22. Belarusian State University, Minsk, 2004.

P. Filzmoser. A multivariate outlier detection method. In P. Filzmoser S. Aivazian and Yu. Kharin, editors, *Proceedings of the Seventh International Conference on Computer Data Analysis and Modeling*, pages 18–22. Belarusian State University, Minsk, 2004.

P. Filzmoser. Robust principal component and factor analysis in the geostatistical treatment of environmental data. *Environmetrics*, 10:363–375, 1999.

P. Filzmoser and K. Hron. Outlier detection for compositional data using robust methods. *Mathematical Geosciences*, 40(3):233–248, 2008a.

doi:http://dx.doi.org/10.1007/s11004-007-9141-5. URL http://www.springerlink.com/content/d662421553216861.

P. Filzmoser and K. Hron. Correlation analysis for compositional data. Research report sm-2008-2, Department of Statistics and Probability Theory, Vienna University of Technology, 2008b. URL http://www.statistik.tuwien.ac.at/forschung/SM/SM-2008-2complete.pdf.

P. Filzmoser, K. Hron, and C. Reimann. PCA for compositional data with outliers. *Environmetrics*, in press, 2008.

M. Fischetti and J.J. Salazar-González. Complementary cell suppression for statistical disclosure control in tabular data with linear constraints. *Journal of the American Statistical Association*, 95:916–928, 2000.

C. Fraley and A.E. Raftery. How many clusters? which clustering method? answers via model-based cluster analysis. *The Computer Journal*, 41(8):578–588, 1998.

L. Franconi and S. Polettini. Individual risk estimation in μ-Argus: a review. In J. In: Domingo-Ferrer, editor, *Privacy in Statistical Databases, Lecture Notes in Computer Science*, pages 262–272. Springer, 2004a.

L. Franconi and S. Polettini. Individual risk estimation in μ-ARGUS: a review. In *Privacy in Statistical Databases. Lecture Notes in Computer Science. Springer*, pages 262–272, 2004b.

H. Fritz and P. Filzmoser. *Plausibility of databases and the relation to imputation methods*. VDM Verlag Dr. Müller, Saarbrücken, 2008.

K.R. Gabriel. The biplot graphic display of matrices with application to principal component analysis. *Biometrika*, 58(3):453–467, 1971.

M. Graf. Swiss earnings structure survey. compositional data in a stratified two-stage sample. Metholology report, isbn: 3-303-00338-6, Swiss Federal Statistical Office, 2006. URL http://www.bfs.admin.ch/bfs/portal/de/index/themen/00/07/blank/02.Document.77975.pdf.

R. Griffin, A. Navarro, and L. Flores-Baez. Disclosure avoidance for the 1990 census. In *Proceedings of the Section on Survey Research Methods*, pages 516–521. American Statistical Association, 1989.

J. Heitzig. The 'jackkife' method: confidentiality protection for complex statistical analyses. In *Joint UNECE/Eurostat work session on statistical data confidentiality, Geneva, Switzerland*, 2005.

J. Heitzig. Using the jackknife method to produce safe plots of microdata. In *Privacy in Statistical Databases. Lecture Notes in Computer Science. Springer*, pages 139–151, 2006.

H. Hofmann and M. Theus. Interactive graphics for visualizing conditional distributions. Unpublished manuscript, 2005.

K. Hron, M. Templ, and P. Filzmoser. Imputation of compositional data using robust methods. Research report sm-2008-4, Department of Statistics and Probability Theory, Vienna University of Technology, 2008a. URL http://www.statistik.tuwien.ac.at/forschung/SM/SM-2008-4complete.pdf.

K. Hron, M. Templ, and P. Filzmoser. Imputation of missing values for compositional data using classical and robust methods. Reserach report sm-2008-5, Department of Statistics and Probability Therory, Vienna University of Technology, 2008b. URL http://www.statistik.tuwien.ac.at/forschung/SM/SM-2008-4complete.pdf.

P.J. Huber. *Robust Statistics*. Wiley and Sons, New York, 1981.

P.J. Huber. Projection pursuit. *Ann. Statist.*, 13:435–525, 1985.

B. Hulliger. Simple and robust estimators for sampling. In *Proceedings of the Survey Research Methods Section*, pages 54–63. American Statistical Association, 1999.

A. Hundepool. The casc project. In *Privacy in Statistical Databases. Lecture Notes in Computer Science. Springer*, pages 199–212, 2004.

A. Hundepool and P-P. de Wolf. Onsite@home: Remote access at statistics netherlands. In *Monographs of official statistics, Work session on statistical data confidentiality*. Eurostat, Luxembourg, 2005.

A. Hundepool, R. Ramaswamy, de Wolf P-P., L. Franconi, S. Giessing, D. Repsilber, J.J. Salazar, C. Castro, G. Merola, and P. Lowthian, 2003. URL http://neon.vb.cbs.nl/casc/TAU.html.

A. Hundepool, A. Van deWetering, Ramaswamy R., L. Franconi, A. Capobianchi, P-P. DeWolf, J. Domingo-Ferrer, V. Torra, R. Brand, and S. Giessing. μ-argus version 3.2 software and users manual, 2005. URL http://neon.vb.cbs.nl/casc.

A. Hundepool, A. Van deWetering, Ramaswamy R., L. Franconi, A. Capobianchi, P-P. DeWolf, J. Domingo-Ferrer, V. Torra, R. Brand, and S. Giessing. μ-argus version 4.1 software and users manual, 2006. URL http://neon.vb.cbs.nl/casc.

A. Hundepool, J. Domingo-Ferrer, L. Franconi, S. Giessing, R. Lenz, J. Longhurst, E. Schulte Nordholt, G. Seri, and P. De Wolf. *Handbook on Statistical Disclosure Control*, 2007a.

A. Hundepool, J. Domingo-Ferrer, L. Franconi, S. Giessing, R. Lenz, J. Longhurst, E. Schulte-Nordholt, G. Seri, and P.-P. De Wolf. Handbook on statistical disclosure control version 1.01, 2007b.

R.L. Iman and W.J. Conover. A distribution-free approach to inducing rank correlation among input variables. *Communications in Statistics*, B11:311–334, 1982.

A.F. Karr, A. Oganian, J.P. Reiter, and Mi-Ja Woo. New measures of data utility. Technical report, SAMSI, data confidentiality group, 2006.

J.P. Kelly, B.L. Golden, and A.A. Assad. Using simulated annealing to solve controlled rounding problems. *Annals of Operations Research*, 2(2):174–190, 1990.

H. Kim, G.H. Golub, and H. Park. Missing value estimation for DNA microarray gene expression data: local least squares imputation bioinformatics. *Bioinformatics*, 21(2):187–198, 2005.

J.J. Kim. A method for limiting disclosure in microdata based on random noise and transformation. In *Proceedings of the Section on Survey Research Methods*, pages 303–308. American Statistical Association, 1986.

J.J. Kim and W.E. Winkler. Masking microdata files. In *Proceedings of the Section on Survey Research Methods*, pages 114–119. American Statistical Association, 1995.

P. Kooiman, L. Willenbourg, and J. Gouweleeuw. A method for disclosure limitation of microdata. Technical report, Research paper 9705, Statistics Netherlands, Voorburg, 2002.

L.Ó. Kovács, G.P. Kovács, J.A. Martín-Fernández, and C. Barceló-Vidal. *Compositional data analysis in the geosciences: From theory to practice*, chapter

Major-oxide compositional discrimination in Cenozoic volcanites of Hungary, pages 145–160. Geological Society, London, 2006. Special Publications 264.

F. Leisch. Sweave, part I: Mixing R and LATEX. *R News*, 2(3):28–31, December 2002a. URL http://CRAN.R-project.org/doc/Rnews/.

F. Leisch. Sweave: Dynamic generation of statistical reports using literate data analysis. In Wolfgang Härdle and Bernd Rönz, editors, *Compstat 2002 – Proceedings in Computational Statistics*, pages 575–580. Physica Verlag, Heidelberg, 2002b. URL http://www.ci.tuwien.ac.at/~leisch/Sweave. ISBN 3-7908-1517-9.

F. Leisch and A.J. Rossini. Reproducible statistical research. *Chance*, 16(2): 46–50, 2003.

G. Li and Z. Chen. Projection-pursuit approach to robust dispersion matrices and principal components: primary theory and monte carlo. *J. Amer. Statist. Ass.*, 80:759–766, 1985.

R.J.A. Little. A test of missing completely at random for multivariate data with missing values. *Journal of the American Statistical Association*, 83(404):1198–1202, 1988.

R.J.A. Little and D.B. Rubin. *Statistical Analysis with Missing Data*. Wiley, New York, 1987.

R.A. Maronna. Robust m-estimators of multivariate location and scatter. *The Annals of Statistics*, 4(1):51–67, 1976.

R.A. Maronna and R.H. Zamar. Robust multivariate estimates for highdimensional datasets. *Technometrics*, 44:307–317, 2002.

J. Martín-Fernández, C. Barceló-Vidal, and V. Pawlowsky-Glahn. Dealing with zeros and missing values in compositional data sets using nonparametric imputation. *Mathematical Geology*, 35(3):253–278, 2003. doi:10.1023/A:1023866030544. URL http://www.springerlink.com/content/ku816485q4264772/.

J.M. Mateo-Sanz, A. Martínez-Ballesté, and J. Domingo-Ferrer. Fast generation of accurate synthetic microdata. In J. In: Domingo-Ferrer, editor, *Privacy in Statistical Databases, Lecture Notes in Computer Science*, pages 298–306. Springer, 2004a.

J.M. Mateo-Sanz, F. Sebe, and J. Domingo-Ferrer. Outlier protection in continuous microdata masking. *Lecture Notes in Computer Science, Vol. Privacy in Statistical Databases, Springer Verlag*, 3050:201–215, 2004b.

J.M. Mateo-Sanz, J. Domingo-Ferrer, and F. Sebé. Probabilistic information loss measures in confidentiality protection of continuous microdata. *Data Mining and Knowledge Discovery*, 11:181–193, 2005.

G. Mateu-Figueras and V. Pawlowsky-Glahn. A critical approach to probability laws in geochemistry. *Mathematical Geosciences*, 40(5):489–502, 2008.

B. Meindl and M. Templ. The anonymisation of the CVTS2 and income tax dataset. an approach using R-package sdcMicro. In *to appear in: Joint UN-ECE/Eurostat Work Session on Statistical Data Confidentiality. Monographs of Official Statistics.*, 2007.

Bernhard Meindl. *sdcTable: statistical disclosure control for tabular data*, 2009. R package version 0.0.2.

G. Merola. Generalized risk measures for tabular data, 2003.

J. Merz, D. Vorgrimler, and M. Zwick. De facto anonymised microdata file on income tax statistics 1998. Technical report, SRI Intl. Tech. Rep., 2005. Nr. 5, 10/2005.

B. Minasny. *Sampling methods for uncertainty analysis*, 2003. Matlab Toolbox for Latin Hypercube Sampling.

R.A. Moore. Controlled data-swapping techniques for masking public use microdata sets. Technical report, U.S. Bureau of the Census, Statistical Research Division Report Series, RR96-04, 1996.

K. Muralidhar and R. Sarathy. Data shuffling- a new masking approach for numerical data. *Management Science*, 52(2):658–ï£¡670, 2006.

K. Muralidhar, R. Parsa, and R. Sarathy. A general additive data perurbation method for database security. *Management Science*, 45:1399–1415, 1999.

K. Muralidhar, R. Sarathy, and R. Dankekar. Why swap when you can shuffle? a comparison of the proximity swap and data shuffle for numeric data. In *Privacy in Statistical Databases. Lecture Notes in Computer Science. Springer*, pages 164–176, 2006.

Samarati P. Protecting respondents' identities in microdata release. *IEEE Transactions on Knowledge and Data Engineering*, 13(6):1010–1027, 2001.

J. Palarea-Albaladejo and J. A. Martín-Fernández. A modified em alr-algorithm for replacing rounded zeros in compositional data sets. *Computer & Geosciences*, 34(8):902–917, 2008.

V. Pawlowsky-Glahn and J.J. Egozcue. Blu estimators and compositional data. *Mathematical Geology*, 34(3):259–274, 2002.

V. Pawlowsky-Glahn, J.J. Egozcue, and J. Tolosana-Delgado. Lecture notes on compositional data analysis. 2005. URL http://diobma.udg.edu:8080/dspace/bitstream/10256.1/297/1/CoDa-book.pdf.

K. Pearson. On lines and planes of closest fit to systems of points in space. *Philosophical Magazine*, 6(2):559–572, 1901.

K. Pearson. Mathematical contributions to the theory of evolution. on a form of spurious correlation which may arise when indices are used in the measurement of organs. In *Proceedings of the Royal Society of London*, volume 60, pages 489–502, 1897.

K. Piker. Geheimhaltung - allgemeiner programmablauf, 1995.

S. Polletini and G. Seri. Guidelines for the protection of social micro-data using individual risk methodology. deliverable no. 1.2-d3, CASC Project, 2004. URL http://neon.vb.cbs.nl/casc/.

Maronna R., Martin D., and Yohai V. *Robust Statistics: Theory and methods*. Wiley, New York, 2006.

R Development Core Team. *R: A Language and Environment for Statistical Computing*. R Foundation for Statistical Computing, Vienna, Austria, 2008a. URL http://www.R-project.org. ISBN 3-900051-07-0.

R Development Core Team. *R Import / Export*. R Foundation for Statistical Computing, Vienna, Austria, 2008b. URL http://cran.r-project.org/doc/manuals/R-data.pdf. ISBN 3-900051-10-0.

R Development Core Team. *The R Langage definition*. R Foundation for Statistical Computing, Vienna, Austria, 2008c. URL http://cran.r-project.org/doc/manuals/R-lang.pdf. ISBN 3-900051-13-5.

C. Reimann, P. Filzmoser, R.G. Garrett, and R. Dutter. *Statistical Data Analysis Explained: Applied Environmental Statistics with R.* Wiley, Chichester, 2008.

R. D. Repsilber. Preservation of confidentiality in aggregated data, 1994.

Y. Rinott. On models for statistical disclosure risk estimation. In *In Proceedings of the joint ECE/Eurostat Work Session on Statistical Data Confidentiality*, pages 275–285, 1990.

Y. Rinott and N. Shlomo. A generalized negative binomial smoothing model for sample disclosure risk estimation. In *Privacy in Statistical Databases. Lecture Notes in Computer Science. Springer*, pages 82–93, 2006.

P. Rousseeuw. Multivariate estimation with high breakdown point. In *Mathematical Statistics and Applications*, pages 283–297. Akademiai Kiado, Budapest, 1985.

P.J. Rousseeuw and K. Van Driessen. Computing lts regression for large data sets. *Estadistica*, 54:163–190, 2002.

Van Driessen K. Rousseeuw P.J. A fast algorithm for the minimum covariance determinant estimator. *Technometrics*, 41:212–223, 1999.

D.B. Rubin. Inference and missing data. *Biometrika*, 63:581–592, 1976.

D.B. Rubin. Discussion of statistical disclosure limitation. *Journal of Official Statistics*, 9(2):461ï£¡–468, 1993.

Oba S., Sato M.A., Takemasa I., Monden M., Matsubara K., and Ishii S. A bayesian missing value estimation method for gene expression expression profile data. *Bioinformatics*, 19(16):2088–2096, 2003.

J.J. Salazar-González. Controlled rounding and cell perturbation: Statistical disclosure limitation methods for tabular data. *Mathematical Programming*, 105: 583–603, 2008a.

J.J. Salazar-González. A unified mathematical programming framework for different statistical disclosure limitation methods. *Oper. Res.*, 53(5):819–829, 2005. ISSN 0030-364X. doi:http://dx.doi.org/10.1287/opre.1040.0202.

J.J. Salazar-González. Statistical confidentiality: Optimization techniques to protect tables. *Computers & Operations Research*, 35:1638–1651, 2008b.

P. Samarati and L. Sweeney. Protecting privacy when disclosing information: k-anonymity and its enforcement through generalization and suppression. Technical report, SRI Intl. Tech. Rep., 1998.

J.L. Schafer. *Analysis of Incomplete Multivariate Data*. Chapman & Hall, London, 1997.

M. Schmid. The effect of single-axis sorting on the estimation of a linear regression., 2006.

M. Scholz, F. Kaplan, C.L. Guy, J. Kopka, and J. Selbig. Non-linear pca: a missing data approach. *Bioinformatics*, 21:3887–3895, 2005.

S. Serneels and T. Verdonck. Principal component analysis for data containing outliers and missing elements. *Computational Statistics & Data Analysis*, 52 (3):1712–1727, 2008.

Statistics Austria. Einkommen, Armut und Lebensbedingungen 2004, Ergebnisse aus EU-SILC 2004, 2006. In German. ISBN 3-902479-59-0.

P. Steel and A. Reznek. Issues in designing a confidential preserving model server. In *Monographs of official statistics, Work session on statistical data confidentiality*. Eurostat, Luxembourg, 2005.

M.L. Stein. Large sample properties of simulations using latin hypercube sampling. *Technometrics*, 29:143–151, 1987.

L. Sweeney. k-anonymity: a model for protecting privacy. *International Journal on Uncertainty, Fuzziness and Knowledge-based Systems*, 10(5):557–570, 2002.

M. Templ. disclosure: Statistical disclosure control methods for the generation of public- and scientific-use files and for the protection of hierarchical tables. R package version 0.97, 2005.

M. Templ. *Privacy in Statistical Databases, Lecture Notes in Computer Science, Vol. 4302*, chapter Software Development for SDC in R, pages 347 – 359. Springer, Berlin Heidelberg, 2006a. doi:http://dx.doi.org/10.1007/11930242_29. URL http://www.springerlink.com/content/4t0h5123v1436342/?p=efe8cfc1405d40debebf9878c83c505b&pi=28. ISBN: 978-3-540-49330-3.

M. Templ. sdcMicro: A new flexible R-package for the generation of anonymised microdata - design issues and new methods. In *to appear in: Joint UNECE/Eurostat Work Session on Statistical Data Confidentiality. Monographs of Official Statistics.*, 2007a.

M. Templ. sdcMicro: A package for statistical disclosure control in R. In *Bulletin of the International Statistical Institute, 56th Session*, 2007b.

M. Templ. *sdcMicro: Statistical Disclosure Control methods for the generation of public- and scientific-use files.*, 2007c. R package version 2.5.1.

M. Templ. Forschungsaktivitäten im bereich statistischer geheimhaltung von mikrodaten. *Statistische Nachrichten*, 1:94–98, 2008a. URL http://www.statistik.at/web_de/services/stat_nachrichten/029481.html#index12.

M. Templ. Making SDC-tools better usable by statistical agencies: Sustainability of the Argus software. Reserach report cs-2008-5, Department of Statistics and Probability Therory, Vienna University of Technology, 2008b. URL http://www.statistik.tuwien.ac.at/forschung/CS/CS-2008-5complete.pdf.

M. Templ. Statistical disclosure control for microdata using the R-package sdcMicro. *Transactions on Data Privacy*, 1(2):67–85, 2008c. URL http://www.tdp.cat/issues/abs.a004a08.php.

M. Templ. Forschungsprojekt AMELI: Regionale Schätzung von Armut und des sozialen Zusammenhaltes in Europa. *Statistische Nachrichten*, 2, 2009. in press.

M. Templ. *sdcMicro. Manual and Package.* Statistics Austria and Vienna University of Technology, Vienna, Austria, 2007d. URL http://cran.r-project.org/src/contrib/Descriptions/sdcMicro.html. *http://cran.r-project.org/src/contrib/Descriptions/sdcMicro.html.*

M. Templ. Experimental R-packages for data disclosure. In *Proceedings of the European Conference on Quality in Survey Statistics*, 2006b.

M. Templ and A. Alfons. *VIM: Visualization and Imputation of Missing Values*, 2008. URL http://cran.r-project.org. R package version 1.2.4.

M. Templ and P. Filzmoser. Visualization of missing values using the R-package VIM. Reserach report cs-2008-1, Department of Statistics and Probability Therory, Vienna University of Technology, 2008. URL http://www.statistik.tuwien.ac.at/forschung/CS/CS-2008-1complete.pdf.

M. Templ and B. Meindl. Robust statistics meets SDC: New disclosure risk measures for continuous microdata masking. *Privacy in Statistical Databases. Lecture Notes in Computer Science. Springer*, 5262:113–126, 2008a. ISBN 978-3-540-87470-6, DOI 10.1007/978-3-540-87471-3_10.

M. Templ and B. Meindl. Robustification of microdata masking methods and the comparison with existing methods. *Privacy in Statistical Databases. Lecture Notes in Computer Science. Springer*, 5262:177–189, 2008b. ISBN 978-3-540-87470-6, DOI 10.1007/978-3-540-87471-3_15.

M. Templ, P. Filzmoser, and C. Reimann. Cluster analysis applied to regional geochemical data: Problems and possibilities. *Applied Geochemistry*, 23(8): 2198–2213, 2008.

M. Templ, P. Filzmoser, and K. Hron. Compositional data using the R ,-package robCompositions. In *In Proceedings of the International Conference on New Techniques and Technologies in Statistics (NTTS 2009), Bruessels*, page 11, 2009a.

M. Templ, K. Hron, and P. Filzmoser. *robCompositions: Robust Estimation for Compositional Data.*, 2009b. R package version 1.2.

T. Templ. *sdcMicro: Statistical Disclosure Control methods for the generation of public- and scientific-use files.*, 2008d. R package version 2.4.7.

M. Theus, H. Hofmann, B. Siegl, and A. Unwin. MANET - Extensions to interactive statistical graphics for missing values. In *In New Techniques and Technologies for Statistics II*, pages 247–259. IOS Press, 1997.

Martin Theus. Interactive data visualization using mondrian. *Journal of Statistical Software*, 7(11):1–9, 2002. ISSN 1548-7660. URL http://www.jstatsoft.org/v07/i11.

D. Ting, S. Fienberg, and M. Trottini. Romm methodology for microdata release. In *Monographs of official statistics, Work session on statistical data confidentiality*. Eurostat, Luxembourg, 2005.

V. Torra, J.M. Abowd, and J. Domingo-Ferrer. Using mahalanobis distance-based record linkage for disclosure risk assessment. In *Privacy in Statistical Databases. Lecture Notes in Computer Science. Springer*, pages 233–242, 2006.

O. Troyanskaya, M. Cantor, G. Sherlock, P. Brown, T. Hastie, R. Tibshirani, D. Botstein, and R. Altman. Missing value estimation methods for dna microarrays. *Bioinformatics*, 17(6):520–525, 2001.

A. Unwin. *Computational Statistics*, chapter REGARDing Geographic Data., pages 315–326. Physica-Verlag, Heidelberg, 1994.

A. Unwin, G. Wills, and J. Haslett. REGARD - Graphical analysis of regional data. In *In Proceedings of the Section on Statistical Graphics, Alexandria*, pages 36–41. American Statistical Association, 1990.

A. Unwin, G. Hawkins, H. Hofmann, and B. Siegl. Interactive graphics for data sets with missing values: MANET. *Journal of Computational and Graphical Statistics*, 5(2):113–122, 1996.

E.J. Wegman. Hyperdimensional data analysis using parallel coordinates. *Journal of the American Statistical Association*, 85:664–675, 1990.

L. Willenborg and T. De Waal. Elements of statistical disclosure control, 2000. ISBN: 0387951210.

G.D. Wyss and K.H. Jorgensen. Sandia's latin hypercube sampling software. Technical report sand98-0210, Sandia National Laboratories, Albuquerque, NM, 1998.

W.E. Yancey, W.E. Winkler, and R.H. Creecy. Disclosure risk assessment in perturbative microdata protection. In *Inference Control in Statistical Databases, Lecture Notes in Computer Science*, pages 49–60. Springer, 2002.

F.W. Young. *ViSta: The Visual Statistics System*. UNC L.L. Thurstone Psychometric Laboratory Research Memorandum 94-1(c), 1996.

F.W. Young, P.M. Valero-Mora, and M. Friendly. *Visual Statistics. Seeing Data with Dynamic Interactive Graphics*. Wiley, New York, 2006.

Bibliography

List of Tables

1.1 Illustration of (primary) sensitive cells in tabular data. (a) shows the median income per age and zip-code. (b) highlights how many individuals correspond to the cells. 17

2.1 An overview of the most important microdata protection methods . 47

2.2 Comparison of different methods regarding a univariate measure of information loss (mean absolute error of medians), one multivariate measure of information loss (mean absolute error of correlations), one utility measure, the risk measure given in Mateo-Sanz et al. [2004b] and a new risk measure weighted with the robust Mahalanobis distance. Further measures were evaluated but not printed in this example. 57

3.1 Investigated methods 64

3.2 Detailed check if the methods can deal with special data configurations, i.e. which methods protect the underlying data well and which methods preserve the data structure. 81

7.1 Implementations of plots related to missing values in selected software. 161

7.2 Explanation of the used variables from the EU-SILC data set. . . . 162

List of Tables

8.1 Estimations of the missing value in cell $[1,3]$ of the expenditures data set (observed value is 147). The considered imputation methods are geometric and arithmetic mean imputation (*gmean* and *mean*), iterative LS and LTS procedure with and without ilr transformation, EM algorithm for alr-transformed (*alr-EM*) and non-transformed (*EM*) data, and $k$nn imputation based on Aitchison and Euclidean distances. *original* corresponds to results for the original data, *outlier 1* is for an outlying observation in both the Aitchison and Euclidean geometries, *outlier 2* is for an outlier only in the Euclidean space. The numbers 1, 2, and 10 are the multiplication factors for generating the outliers. **182**

List of Figures

1.1	Risk-Utility confidentiality map (RU map).	2
2.1	Computation time for calculating the frequency counts for the μ-Argus test data set with 4000 observations.	41
2.2	Survey on certain procedures in package *sdcMicro* and their relationship.	42
2.3	Individual risk. In the upper region of the figure you will see a helpful slider which is directly linked with the graphics.	53
2.4	Individual risk after the anonymization of the μ-Argus test data set.	56
3.1	LEFT: The original bivariate normal distributed data (circles) and the shuffled data (crosses) do have a quite similar behavior. RIGHT: The original data consists of bivariate normal distributed data (non-outliers) plus a shifted outlier group and the shuffled data (crosses) show a quite dissimilar behavior.	75
3.2	LEFT: The original bivariate normal distributed data (circles) and the shuffled data (crosses) do have a quite similar behavior. RIGHT: The original data consists of bivariate normal distributed data (non-outliers) plus a shifted outlier group and the shuffled data (crosses) also show a quite similar behavior.	76

List of Figures

3.3 TOP LEFT: IL1s information loss measure versus the amount of shifted outliers. TOP RIGHT: disclosure risk (Mateo-Sanz et al. [2004b]) versus the amount of outliers in the generated data sets. BOTTOM LEFT: Information loss based on differences of the eigenvalues of the robust covariances between original and perturbed data versus the amount of outliers. BOTTOM RIGHT: weighted disclosure risk based on robust Mahalanobis distances versus the amount of outliers. 82

4.1 Illustration of the concept of Mahalanobis distances and robust Mahalanobis distances on a simply 2-dimensional example. LEFT: Tolerance ellipse (95 %) and "outlier detection" using Mahalanobis distances. RIGHT: Tollerance ellipse (95 %) and outlier detection using robust Mahalanobis distances. 89

4.2 Original observations and the corresponding masked observations (perturbed by adding additive noise). In the bottom right graphic small additional regions are plotted around the masked values for *RMDID2* procedure. 97

4.3 Influence of parameter k on different distance-based disclosure risk measures. 98

4.4 Effects of shifted outliers (0 till 40 percent in 2.5 percent steps) on some protection methods based on *SDID* measure. 99

4.5 Effects of outliers on some protection methods based on *RMDID2* measure. 100

4.6 Comparison of *SDID* and *RMDID2* measure under different contaminations. 101

4.7 Disclosure risk evaluated for every observation. 101

5.1 Data utility resulting from different perturbation methods 114

List of Figures

5.2 Disclosure risk resulting from different perturbation methods 114
5.3 Comparison plot of original and microaggregated data via sorting on the first classical principal component. 116
5.4 Comparison plot of original and microaggregated data via sorting on the first robust principal component with projection pursuit in each cluster. 116
6.1 individual risk (left) and empirical distribution (right) in original data. 126
6.2 individual risk (left) and empirical distribution (right) in anonymised data. 128
6.3 individual risk (left) and empirical distribution (right) in original data. 131
6.4 individual risk (left) and empirical distribution (right) in anonymised data. 133
7.1 Simulated bivariate data set. LEFT: Complete data. MIDDLE: Red points are chosen as missing in y, depending on the value of y (MNAR). RIGHT: Information is only available for x-values in practice. 141
7.2 Simulated bivariate data set. LEFT: Complete data. MIDDLE: Red points are chosen as missing in y, depending on the value of y (MNAR). RIGHT: Information is only available for x-values in practice. 142
7.3 Number of missing values for a subsample of the EU-SILC data from Statistics Austria. LEFT: Barplot of the number of missing values in each variable. RIGHT: *Aggregation plot* showing all existing combinations of missing (red) and non-missing (blue) values in the observations, and the corresponding frequencies. 146
7.4 Matrix plot of a subset of the EU-SILC data, sorted by variable *pek_n*. 147
7.5 Histogram (left) and spinogram (right) of *age* with color coding for missing (red) and available (blue) data in variable *py010n* (employee cash or near cash income). 149

List of Figures

7.6 Scatterplot of *pek_n* (net income) and *py130n* (disability benefits), both log-transformed, with information about missing values in the plot margins. 151

7.7 Parallel coordinate plot for a subset of the EU-SILC data. Red lines indicate missing values in variable *py050n* (cash benefits or losses from self-employment). 153

7.8 The transformed values of variable *pek_n* (net income) are grouped according to missingness in several variables of the EU-SILC data and presented in parallel boxplots. 154

7.9 Map of the Kola region. Missings in *As* or *Bi* show a spatial dependency. 156

7.10 Map of the nine political regions of Austria. Regions with a higher proportion of missing values (see the percentages in each region) in variable *py050n* (cash benefits or losses from self-employment) receive a higher portion of red. Further information is provided when clicking on a region in the map. 157

7.11 The **VIM** GUI. 159

8.1 Left plot: Two-part compositional data without the constraint of constant sum. The points could be varied along the lines from the origin without changing the ratio of the compositional parts. Right plot: According to the relative scale, the points close to the boundary are more different than the central points, although the Euclidean distances are the same. The Aitchison distance accounts for this fact. 168

List of Figures   *List of Figures*

8.2 Left plot: Two-part compositional data consisting of three groups, and their projection on the line indicating constant sum 1. While the relative information of the groups with symbols ○ and △ is similar, the data points of the group with symbol + contain very different information. Right plot: In the upper part the ilr transformed original data are shown. The lower plot shows the ilr transformed data with constant sum constraint. This demonstrates that the constant sum constraint does not change the ilr transformed data. **171**

8.3 Simulated data set with 5 points from *outlier group 1* (symbol +) and 5 points from *outlier group 2* (symbol △). Left plot: 3-part compositions plotted in the ternary diagram; right plot: data after ilr transformation. **185**

8.4 A set of four subfigures. **187**

8.5 A set of four subfigures. **190**

9.1 Simulated data set with 5 points from *outlier group 1* (symbol) and 5 points from *outlier group 2* (symbol △). Left plot: 3-part compositions shown in the ternary diagram; right plot: data after ilr transformation. **199**

9.2 Boxplot comparison of the estimated column-wise geometric means of 1000 bootstrap replicates of the original data set and the data sets where missing values were imputated with different methods. The red line is the column-wise geometric mean of the original data. Outliers are excluded in the computation of the geometric means. **203**

9.3 Multiple Scatterplot to highlight imputed observations. **204**

9.4 Ternary diagram with special plotting symbols for highlighting imputed parts of the compositional data. **205**

233

Index

Imputation, 27–29, 135–205
    D-part composition, 165, 194
    k-nearest neighbor imputation, 28, 174, 196
    k-nearest neighbor imputation due to compositions, 175
    additive logratio transformation, 170
    adjustment factors for $k$nn, 175
    aggregation plot, 144
    Aitchison distance, 169, 196
    Aitchison geometry, 195
    alr-EM algorithm, 173
    bootstrap, 202
    compositional data, 194
    Compositional error variance, 184
    compositional outlier, 169
    compositional parts, 167
    constant sum contraint, 166
    Difference in covariance structure, 185
    difference in variations, 201
    EM algorithm, 28, 165
    equivalence class, 167
    EU-SILC data, 144
    geometric mean imputation, 173
    ggobi software, 138
    histogram with missing values, 147
    household expenditure data, 179
    inverse ilr transformation, 177
    isometric log-ratio transformation, 197
    isometric logratio transformation, 169, 176
    isometric property, 197
    isometry, 170
    item non-response, 27
    iterative model based imputation, 177
    log-ratio transformation, 195

Manet software, 138
mapping with missing values, 152
MAR, 139
matrix plot, 145
MCAR, 139
MNAR, 139
model-based iterative imputation, 197
multiple scatterplot for imputed values, 203
outlier, 29, 183
parallel boxplots for missing values, 152
parallel coordinate plot for imputed values, 203
parallel coordinate plot with missing values, 150
R package robCompositions, 197
R-package VIM, 138, 155
relative Aitchison distance, 201
robust regression, 179
scatterplot with missing values, 148
sequential binary partitioning, 172
simplex, 166
ternary diagram, 184, 204

uncertainty of imputation, 202
unit non-response, 27

SDC, 1–27, 31–133
  macrodata, 14–27, 116–118
    τ-Argus, 16
    p-rule, 15
    (n,k)-rule, 15
    attacker problem, 17, 19, 118
    cost function, 16
    disclosure risk, 15
    EU-aggregates, 118
    hierarchical tables, 16, 24
    hitas approach, 16
    hypercube method, 116
    integer linear program, 20, 23
    linear program, 17, 117
    lower and upper protection level, 22
    multi-dimensional hierarchical table, 21
    primary cell suppression, 15, 116
    R package disclosure, 15, 18
    R package sdcTable, 17
    secondary cell suppression, 15, 22, 26, 116
    table, 14

INDEX

microdata, 1, 14, 31–96, 121–133
μ-Argus, iii, ix, 34
k-anonymity, 13, 43, 55
adding additive noise, 67
adding correlated noise, 67
adding noise, 44, 66
Austrian income tax data, 128
Benedetti-Franconi model, 10
biplots, 111
blanking and imputation, 71
categorical key variables, 5, 41, 125, 129
confidential preserving model servers, 33
confidential preserving model servers, 106
continuing vocational training survey, 124
copula based gadp, 70
data masking, 34
data utility, 75
disclosure risk, 46
distance-based disclosure risk, 72, 85, 94
distance-based disclosure risk, robust, 91
frequency counts, 5, 51

gadp, 44
global recoding, 12, 41, 52, 125, 131
global risk measures, 5
IL1 information loss measure, 113
inclusion probability, 4
individual ranking, 68
individual risk, 5, 10, 12, 13, 41, 51, 55, 126, 131
information loss, 57, 71
IPSO synthetic data generator, 70
k-anonymity, 132
latin hypercube sampling, 71
local suppression, 12, 43, 55, 126, 131
Mahalanobis distance, 88
Mahalanobis distance, robust, 89
mcd estimator, 45
mdav algorithm, 46
microaggregation, 43, 55, 68, 108
model-based clustering, 108
outlier, 3, 44, 46, 67, 86, 112, 115
output checking, 33

perturbation, 14
pram, 43, 58
projection pursuit, 110
public use files, 3
R package sdcMicro, iii, ix, 7, 11–13, 38, 50, 85, 123
rank swapping, 68
remote access, 32
remote execution, 32
robust gadp, 73
robust microaggregation, 45, 58
robust shuffling, 44, 73
romm, 44, 67
ru-map, 2
S4 class system, 111
sampling weights, 6
scientific use files, 3, 122
shuffling, 44, 69
standardised data sets, 4, 122
structural business statistics data, 113
sub-sampling, 129
Sweave, 105
synthetic data generation, 70
task related data sets, 4
uniqueness, 5, 14

Die VDM Verlagsservicegesellschaft sucht für wissenschaftliche Verlage abgeschlossene und herausragende

Dissertationen, Habilitationen, Diplomarbeiten, Master Theses, Magisterarbeiten usw.

für die kostenlose Publikation als Fachbuch.

Sie verfügen über eine Arbeit, die hohen inhaltlichen und formalen Ansprüchen genügt, und haben Interesse an einer honorarvergüteten Publikation?

Dann senden Sie bitte erste Informationen über sich und Ihre Arbeit per Email an *info@vdm-vsg.de*.

Sie erhalten kurzfristig unser Feedback!

VDM Verlagsservicegesellschaft mbH
Dudweiler Landstr. 99
D - 66123 Saarbrücken

Telefon +49 681 3720 174
Fax +49 681 3720 1749

www.vdm-vsg.de

Die VDM Verlagsservicegesellschaft mbH vertritt

Printed by Books on Demand GmbH, Norderstedt / Germany